Terror of Global Warming

Is it a Hoax?

By Steve Preston

1ˢᵗ Edition

Table of Contents

Introduction

Global warming, Global Warming! Quick get rid of you underarm spray as it might let hydrocarbons into the air! Even with 50% of our Electrical energy coming from coal, we need to eliminate the use of burning coal or all our lands will dry up! Now build electric cars to eliminate the destructive use of gasoline. Wait a minute! I'm back to coal again making electricity. Greenhouse gases became the villains of the 21st century as they pushed us closer to destruction.

While many indicate they are climatologists, the word is nebulous and easily mischaracterized. There are climatologists claims from individuals with degrees in mathematics, atmospheric sciences, meteorology, oceanography, physics, chemistry, geography, geology, astrophysics, statistics, astronomy, engineering, earth sciences, environmental sciences, and more specific fields within those already mentioned. It seems all you need to do is investigate climate and that is what you are. Therefore I am presenting this book as a climatologist with degrees in Engineering and specific knowledge of atmospheric absorption. OK! Most of the climatologist claims are bogus, but that's what has happened since Al Gore the Nobel Prize winning Climatologist in 2007.

Scientists like Al Gore told us *the Earth is dying because of man* and we gave him the Nobel Peace Prize even though he only brought loss of peace and an abundance of fear. He went to accept the tribute in his massive, private, fossil fuel guzzling jet and praised how we were turning a corner with horribly inefficient windmills, minimally effective solar panels, making plants artificially grow faster with DNA modification to reduce the need for fertilizers, and on and on. This is such an important issue; we must make a determination of truth. Unfortunately, there are many levels of truth in a world of what we can call vain truth. Our vanity establishes a truth built around what we WANT TO BELIEVE rather than an absolute truth that runs our reality.

We know something is happening, but what it is and how it will affect us needs to be understood without shouting unfounded claims to support some pet project or money scheme. While our temperature has not been affected, greenhouse gas concentration SEEMS to be steadily going up for the last 200 years as shown next. While some tell you reasons, no one actually knows, for sure why it is happening and what effect, if any, this increase will have. It should be noted that while these increases have been steady, the gases identified here make a **tiny** portion of our atmosphere. Instead of logical debate and scientific research into this anomaly, we only hear ranting.

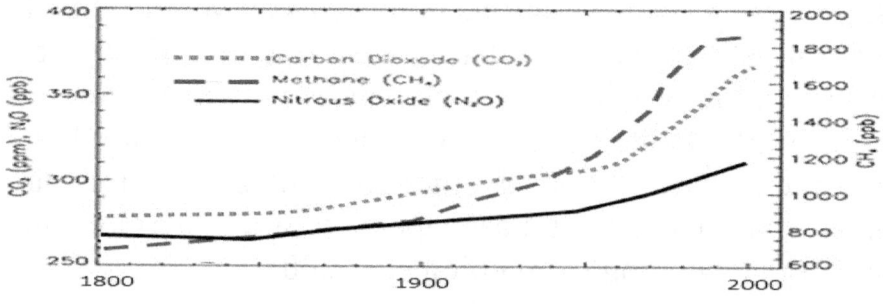

What Are You Hearing?

I don't claim to know the details of Global temperature stability or lack thereof, but I surely can show others where the current claims are unfounded, or blown out of proportions, or directed towards human interference rather than natural phenomenon. We need to keep an open mind about all this as our Earth is a delicate and important part of our existence and if there is something we can do to protect it we should. Here are a few of the things we hear. Later, I will get into these and others in detail but right now let's just get a reasonable list. I'll try not to comment too much.

- **Runaway CO_2 and Methane levels are going to destroy us**- Sorry, but this one is going to make you mad. Supposedly, it is well known that massive increases in greenhouse gasses will cause thermal runaway. Before a couple hundred years ago CO_2, NO_2, and Methane levels remained more constant than the last 200 year. Something may have happened in the 18th century to trigger this extended release, or the solar system may have shifted, or something bad as shown in the graph to the left. To .the right we see the temperature and CO_2 levels over the past 20 thousand years. Just as scary as scary can be. The CO_2 levels are destroying the earth.----so it seems.

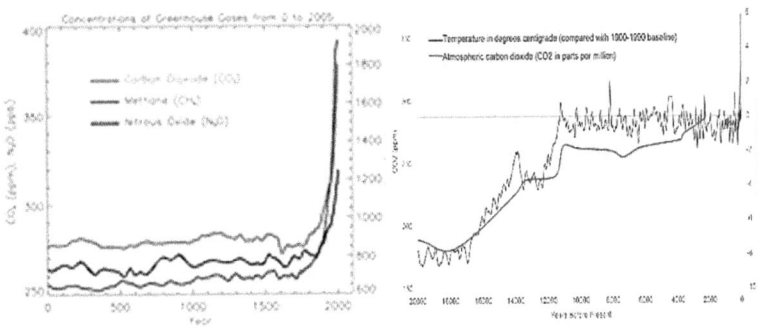

7

I think I'll wait a little before I get into the subtrafuge. Let me just give you a sneak peak. Those prevous graphs were made from a group of charts. The main ones of concern are the CO_2 levels packed into ice core samples from Antartica [up through 1996] and the aresol CO_2 from Hawaii. Knowing that only a portion of CO_2 will be trapped in the Ice, one can quickly see the dread is manufactured by some pretty evil people. Below is is a much more accurate set of graphs.

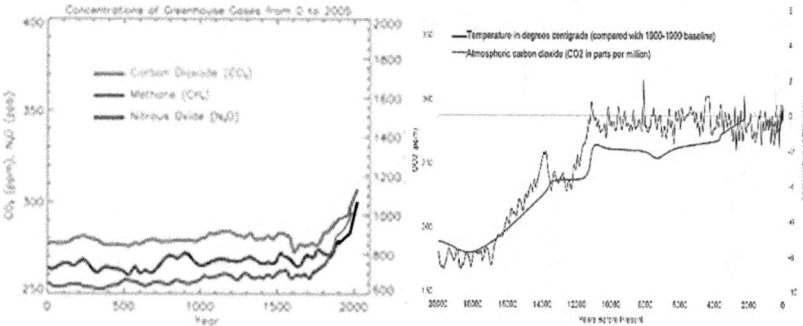

I'll expain all this later and show you just what they did. It is pretty horrible and it was NO ACCIDENT.

- **Global Temperatures are [supposedly] Skyrocketing-** You here, it is "well known" that every year it gets hotter. Never mind the snow blizzards and such, just look at the graphs. This was "compiled by University of East Anglia. It shows about ½ degree over a 160 years, but it looks really bad. Later we will see that the data was doctered to make this graph, but people still are being shown this to get them to support solar and wind power industries.

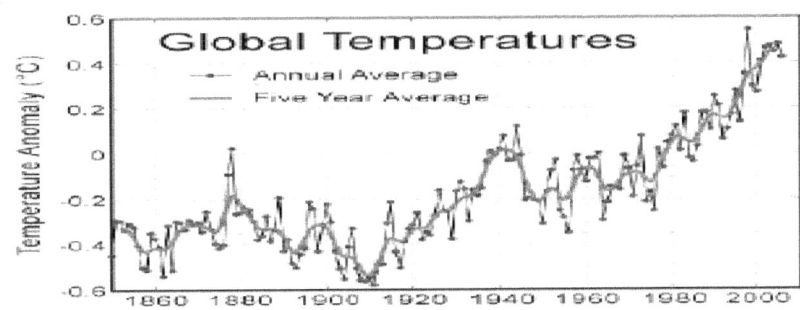

This one is going to make you angry as well as the entire chart has been masterfully altered substantially and no one went to prison even after the act was found out and they had to admit the attempt at driving fear into the entire world. Right now, let me go on or I'll never get through the introduction.

- **Computer Modeled Destruction**- NASA "climate" scientist, James Hansen, modeled the changes and told us that the increases in CO_2 and methane are *"...equivalent to exploding 400,000 Hiroshima atomic bombs per day 365 days per year. That's how much extra energy Earth is gaining each day."* [That's four per second] While *all models have failed* [and our earth is much safer], trying to make them work has also been difficult as many countries, Weather Bureaus, Climate Research Centers, Universities, and the IPPC (United Nations) have put more trust in these theoretical models than actual published measured Satellite and Balloon data. The graph below shows the NATO model the top line is if we keep using dirty stuff and the third line is if we go back to horse and carriage. The bottom 2 lines are actual temperatures from satellites showing how bad the prediction modeling really was as we are doing much better than horse and buggy.

9

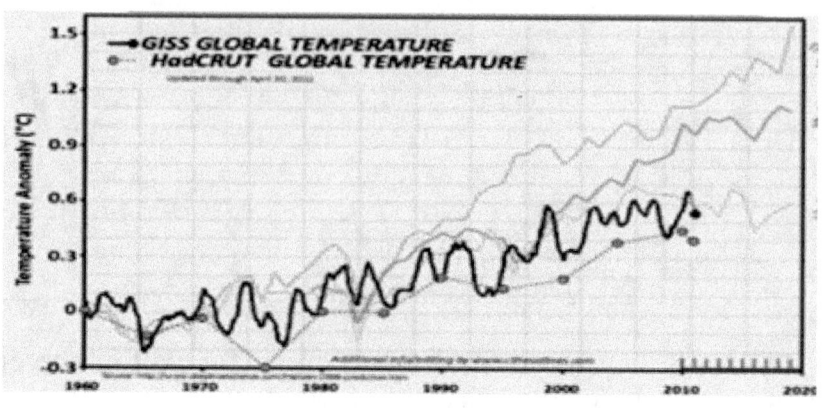

- **Worse Than That**- He continues, *"Even if we stopped burning all fossil fuels tomorrow, our past CO_2 emissions will continue to add hundreds of thousands of A-bombs worth of energy each day for years. Global temperatures will continue to rise and extreme weather events will continue to get more extreme for decades. It will take nearly a century for enough of this CO_2 to be removed from our atmosphere to return our planet back to the temperatures and weather energy levels we are experiencing now."* He even made computer models to show the destruction.

- **Coal Is Our Enemy**- James Hansen continues, "<u>Coal is the single greatest threat to civilization and all life</u> on our planet. From his words and the many "Me-too" quasi-scientists, our massively important coal industry is in shambles. We are so afraid our government has stopped 150 proposed coal plants and "retired" 110 existing coal plants. Now the US EPA passed new clean air regulations that will cause the rest of the coal plants to shut down and impossible for new ones to be build forcing our valuable resource to be thrown away like garbage. [Coal use in the US plunged 13 percent in the

10

last six years as those owning stock in natural gas cheer and renewable energy sources are bankrolled as shown below-numbers are in millions of dollars.]

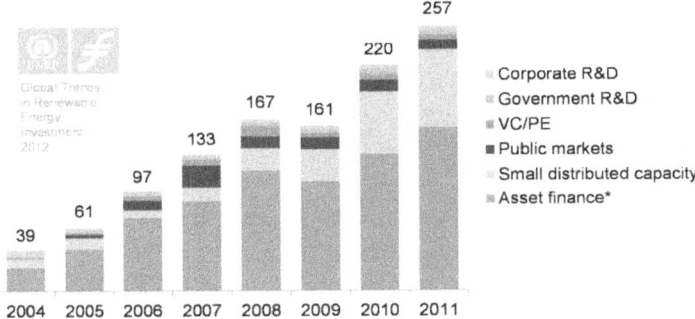

- **More About Coal-** Coal combustion creates 40 percent of electricity worldwide but also is [supposedly] responsible for "*30 percent of total anthropogenic [Man-caused] carbon dioxide emissions worldwide, and 72 percent of CO_2 emissions from global power generation*". As we are causing global warming, we can CERTAINLY see renewable energy sources like wind or solar should be more cost-effective in the long run. Quickly, the United States made plans to eliminate nasty Coal and its working as shown below. Don't worry about those in the coal industries, they can find other work and don't worry that the cost of electricity will skyrocket as soon as coal is not an alternative. A more correct true will amaze you.

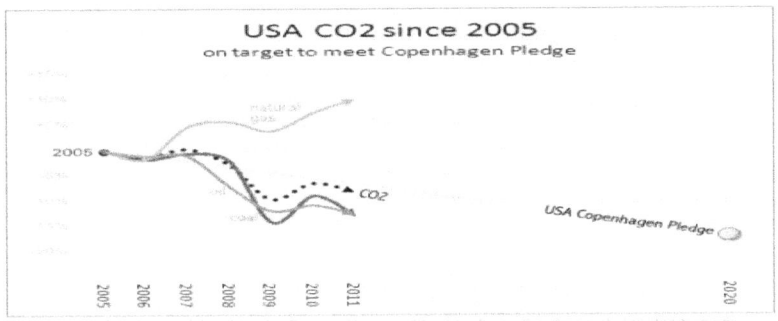

11

- **Melting Ice Caps are going to destroy us-** The melting ice has caused sea levels in the north to rise. For the first time in hundreds of years, ships can pass through the fabled Northwest Passage above North America. Over 100 million people living in coastal regions will be displaced by just a one-yard rise in sea levels. This is also is questionable and we need to examine more details.

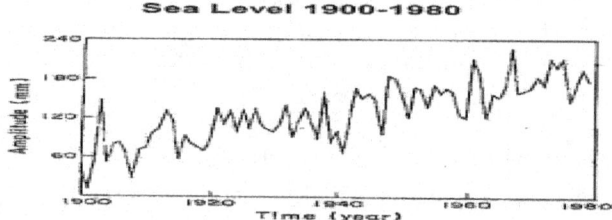

- **Loss of Glacier area in Alaska-** One thing that is for sure, the size of many glaciers in Alaska are getting smaller as the mean temperature in Alaska is higher. The famous McCarty Glacier picture is pasted everywhere showing this fearful sight.

While it is known some areas of the Earth are getting warmer, just as many are gettign cooler as we will see.

- **Extinctions are occurring**-The golden toad was the first species to go extinct because of climate change and habitat loss.

- **Complete Scientific Agreement** -Among climate scientists, 97 percent agree [Ha!] that human-caused [Ha! again] climate change is happening here and now. The sooner we act to slow the rate of climate change, the lower the risk and cost for future generations. [Sorry for the laughter. I'll do better and there is no question that our climate is changing. The question we will investigate is WHY and what we should do about it.]

- **Complete Agreement by Population**- Only 37 percent of Americans believe that global warming is a hoax and 64 percent don't believe that climate change will seriously affect their way of life. We can hope this percentage is going way down after the "doctored data" from NOAA came out, but look at the gallop pole from 2012 showing most believe the global environment changes are manmade. Hopefully I can do my part at fixing some of the wrongs done to everyone by some

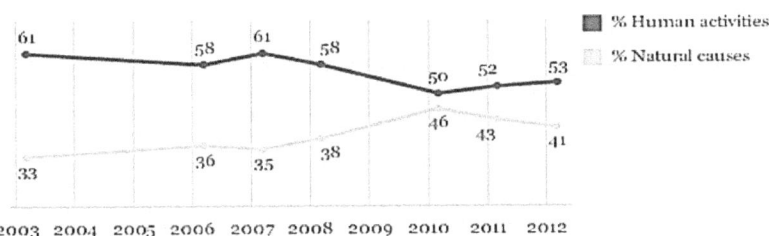

Worse than that; when we look at what happened in 2007 when Al Gore made fake global warming popular, we see that everyone's opinion began changing to global warming away from just about everything else. Every year the fear gets worse and worse.

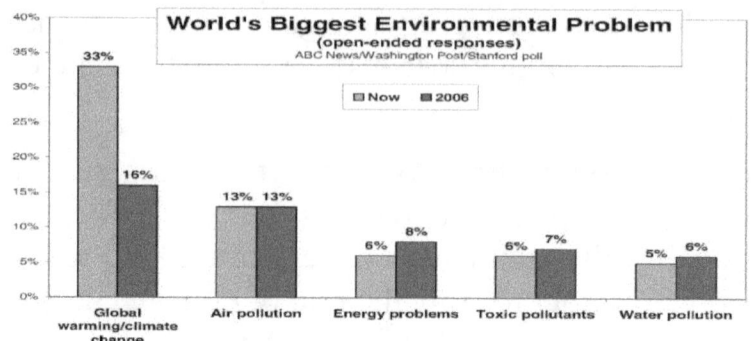

Today by some studies over 90% of the population including those who should [and probably do] know better say manmade global warming is their greatest fear as shown below.

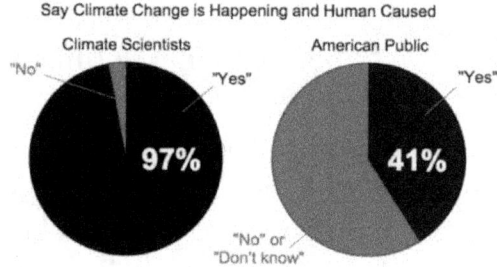

- **Methane and Flatulence-** We are told cows produce more methane than the oil industry does: 20 percent of U.S. methane gas emissions is produced by farmed cattle burps and farts. [No telling what dinosaur flatulence did.]

- **Drought-** They tell us, because of global warming, of the land in California, 99.84 percent is experiencing drought.

- **Huge Raging Storms** – According to the chart following, since 1992 until now North Atlantic storms have increased by 50% because of global warming.

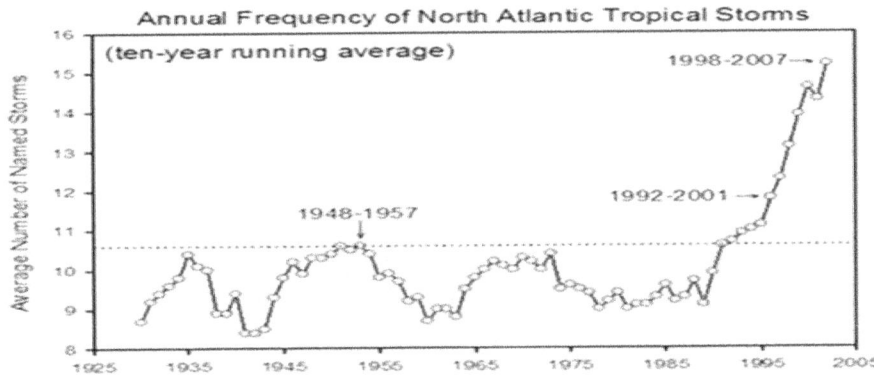

Annual Frequency of North Atlantic Tropical Storms (ten-year running average)

- **Chinese Polluters**- While the U.S. is trying to reduce their greenhouse gases, China plans to construct one coal-based electrical power station each week for the next 10 years. We are getting rid of the nasty things and China is simply laughing.

- **Problem is Well Known**-In 1896, Swedish scientist Svante Arrhenius was the first to claim that burning fossil fuels <u>may</u> eventually result in global warming.

- **High Cost of Global Warming**- Climate change costs the U.S. over $100 billion each year.

Global investment in renewable power and fuels increased 17% to a new record of $257 billion in 2011. Developing economies made up 35% of this total investment, compared to 65% for developed economies. The US had a 57% leap in its outlays to $51 billion. India increased its efforts by 62% to $12 billion in 2011 alone.

- **Loss in our Seas**-The world lost about 16 percent of all coral reefs in 1998, the second hottest year on record due to greenhouse gasses [according to doctored data]. We have caused this!! Since the beginning of the Industrial

Revolution, the acidity of surface ocean waters has increased by about 30 percent. Massive Thermal Runaway will hit us very soon as ozone is disappearing, our surface temperature is unlivable and acid in the oceans is killing everything as shown below. The problem is they are not showing the truth.

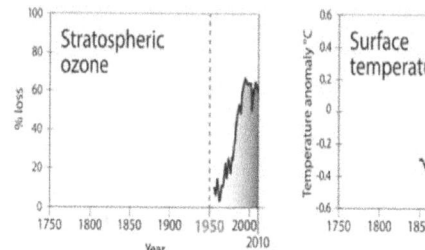
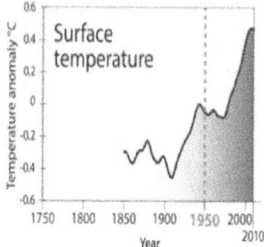
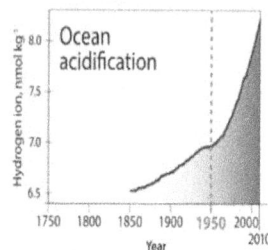

Ice is all Leaving

If that isn't enough, we hear the Arctic is so hot now the Northwest Passage is navigable for the first time in hundreds of years. Antarctica and Greenland Ice accumulations are greatly reduced from previous times. Certainly, we should be able to determine if the northern regions are getting colder simply by testing the Ice Mass or the Sea Ice extent. The normal graphs shown, paint a dismal picture. Taken at the tip of South America, the Antarctic Ice Mass seems to be going away quickly. The graph on the right shows huge reductions in Antarctic mass over a 6 year period..

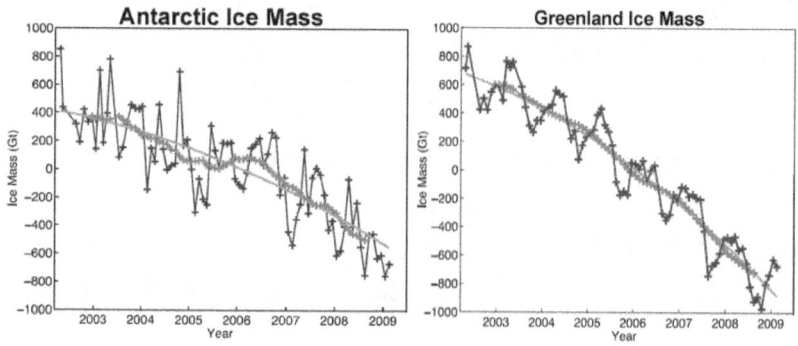

The graph on the preceding right shows almost the same catastrophe in Greenland and the graph below show the ocean heat is rising as fast as a fish can swim.

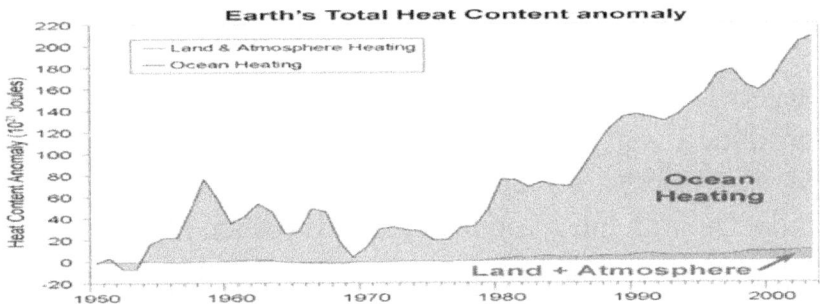

Wait just a minute!!! This doesn't sound right. Satellites show the opposite. Fortunately, some of this stuff is exaggerated, manipulated, falsified, introduced by unqualified quasi-scientists, and many details are buried to keep the world populations supporting the green-gas elimination industries and even to get a Nobel Peace Prize or two. I've tried to ignore the whole greenhouse gas catastrophe for some time now, but I'm getting scared as we would rather weaken our nation than look at facts rather than smiley faced, misdirecting, conniving, sinister, fear-mongers that use tricks to make their particular "green" company flourish rather than helping anyone. This book simply will go through the data and let you make up your mind. We'll look at previous extinction times and determine what happened   We will look at Ice Cores, hot spots traveling in the Pacific, sun spots, the very hot planet Venus, the other weather sampling data that has been truncated to make it appear to be bringing us to calamity, nuclear decay timing, cow and dinosaur flatulence, destruction models, and even cross compared prophecies that help us understand how our Earth behaves and how we have been lied to.

What we will find is the following:

- We have been lied to over and over and over again and the fear of Global Warming has been greatly expanded with this false data.

- Even after the culprits were found out, the data produced is still the basis for establishing terror to allow for massive government support of inappropriate industries at the expense of many other more important tasks to keep our population safe and able to work.

- Huge quantities of massive dinosaurs expelled flatulence and our Earth continued.

- A huge Meteor hit the earth and caused millions of tons of magma to be expelled without our earth atmosphere being affected over the long term.

- Many times in our ancient history, massive civilization rose and fell without much of an impact on our atmosphere.

- Nuclear wars nearly destroyed the people of the planet and no significant change in temperature was noted.

- Climatologists tie different types of data together without regard for reason to make horrible claims that expand the fear of the destruction of our planet.

- Ancient seers of the future; Nostradamus, Mother Shipton, John and Estras from the Bible, and many others indicate that within a very few years from now, there will be a massive, externally induced shift in the environment, but there is no mention of heating. Instead we keep seeing the signs of our next Ice Age.

To start off, I'm going back to the worst destruction man may have ever witnessed. Yes; I'm talking about the complete destruction of the super continent of Prestonia leaving Pangea as the only remaining super continent. The time is near the Permian-Triassic Extinction. There is no Pacific Ocean and no moon. No one was worried about global warming, but there was a catastrophe coming just the same. As we take this second look at our history, please understand the Earth did not melt as billions of people came and went through war and devastation not associated with global warming.

Some of this will seem strange to you as the Global warming misdirection has also been used to rewrite history over and over again. Once we get a better understanding of a more probable history, we'll look again at the notion of global warming with our eyes more open. What we find is that we have been lied to and there is substantial proof, but no one has had to pay for the extortion of fear that has made many rich and hurt us all.

This next section deals with our earth up until the end of the Pleistocene and into the Holocene before the United States became a nation.

The middle section deals with the debate concerning manmade global warming and the levels vested partners will go to until their ideas are considered true, no matter what they know to be the consequences. Also relevant details are brought up concerning the other side of the argument.

The last section deals with impending changes, what we can do about it, if anything, and how these changes will affect us in the near future until the end of time.

Let me start way back in time as a substantial greenhouse gas producer was beginning to take shape. I'm talking about huge Dinosaurs. To get huge, the Earth had to get smaller.

Mars Destroys Prestonia

Back then, the positions of Mars, Venus and Earth were not where they are today; in fact, most of the revolutions of planets surrounding our sun wobbled a little which caused issue for a time. Gravitational pulls from both of our nearby planets have caused great damage at one time or another and this would be one of them. If you have ever wondered where the massive mountain ranges came from, I'll give you a hint. The 2 major ones "the mountain string including the Himalayas" and the string or mountains that start at Antarctica, come up the west coasts of South and North America and travels down the other side of the Pacific in Asia, both were NOT made by plate tectonic as you were taught. While I don't want to get into a science study of the very real repositioning of plates of the Earth's mantle, let me just say that those telling you that plates were pushed in opposite directions simultaneously to build these extremely long ranges and were pushing the crust up miles into the air were not telling you the truth. Instead these two ranges, at least, were pulled up from opposing gravitational forces of a nearby massive object. It is believed that about 400 thousand years ago, Mars orbit brought it fairly close to the earth for a short time. The gravitational pulls of both planets tugged on the planet exterior surfaces. Along the equatorial regions of both planets massive mountain ranges pull up and perforated the exposed land. About half of the Martian surface was yanked away and well over 1/3 of the Earth's surface pulled away as well. A portion got caught up as a satellite we call the moon, but most chunks mixed together and were swept out in space until the sun's gravitational pull began the orbit in the area known as the Asteroid belt. Please see the graphic I made next. The chucks that left, had once been the **Super Continent called "Prestonia" by some.** [I'm embarrassed to say I'm the only one calling this supercontinent by that name, but may ignore its existence altogether, so I had to call it something.] The remaining land "Pangea", almost immediately began to split apart to fill in the hole. This is the type of catastrophe that could cause our planet to lose its hemostasis and blast the earth temperature into a spiral that would destroy mankind. No hydrocarbons were used and the greenhouse effect was not able to convert the Earth into a ball of fire like Venus.

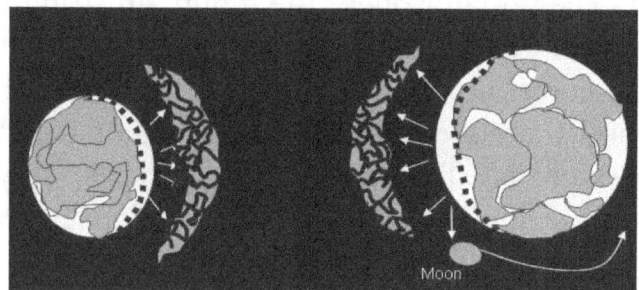

Animals Died Without a Greenhouse Effect

If you were one of those who thought that the super continent of Pangea was on one side of the Earth and <u>nothing</u> was on the other before this major event in Earth's temperature, weather, and life model was modified, you would have a very wobbly earth. While it is very difficult to determine when all this happened, we can be pretty sure massive extinctions occurred.

I Need to Bring up Something Important

After the Mars incident happened, The Earth did something somewhat strange. Like a figure skater doing a spin, as the

Earth got smaller, it began spinning on its axis much faster. This faster spin, made the gravity of the Earth lower so animals had to compensate.......... They got bigger. During the Mesozoic period following the creation of the Pacific Ocean, Massive dinosaurs erupted seemingly overnight. I know "scientists" have told you the Tyrannosaurus Rex could not run for fear of falling on his massive head, the Diplodocus and Pachycephalosaurus could not lift their 50 foot necks as blood could not flow, and the Pterodactyl could not fly as it was simply too heavy, but this also is a lie as all would have died if they could not run, fly and raise their heads to eat. General images of these anomalies are shown below.

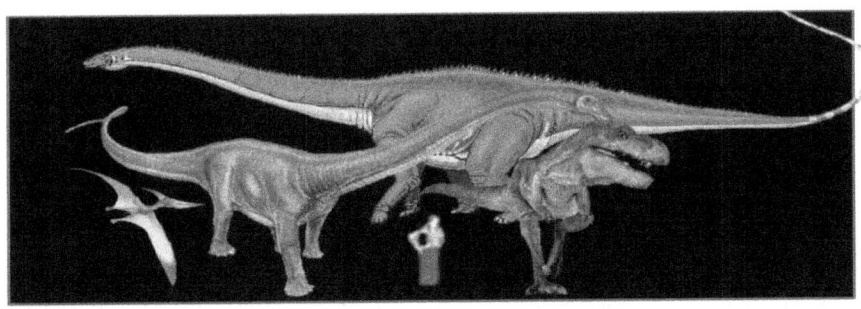

As the Earth rotation increase, the gravity got lower so the animals could survive and thrive. This would have been a stinky time as the dinosaur flatulence was awful. Those studying Cattle flatulence have a terrible job, for sure, but back them the hydro carbon production of these huge animals was tremendous.------No! The earth did not go into a global warming catastrophe, but I don't think about that as we must continue to test the damaging output from cows as it puts people to work fearing the hydrocarbon byproducts of burning coal and cattle rear-ends.

Oh Boy!

I know all this stuff sounds foreign to many who are reading and I will briefly provide some of the details, but this is simply background for our detailed analysis of global warming I will be concentrating on here. There are many books on the things I have addressed and I have written books on these subjects as well. Before we go on, I have to bring up something else you should have been told about in school, but somehow they forgot to tell you. We know people lived with the dinosaurs. There is NO DOUBT as thousands of pieces of physical evidence, historical reference, and religious documents **all** tell us exactly the same thing. Like the animals; people living during this time grew HUGE. Some called them *Titans*, the Book of Genesis called them *"Giants of Old"* [Genesis 6].

Abominable Animals

Living during the Mesozoic Period, we find their remains around the world and what we find shows them to have gained high levels of expertise in genetics, physics, manufacturing, medicine, electricity, and nuclear development. For those who read the Biblical texts and wonder why God told Noah that most of the animals were "abominations or unclean" while only a very small subset of the animals living during his time were "clean". Some tried to tell you an unclean abomination simply was an animal Noah was not to eat, but they are simply mixing up those traveling with Moses and Noah, who was living during the Pleistocene Age. The reason most animals were hated by God is that he did not make them and ancient texts of the Sumerians, the Brazilian Mongulala, and the Jewish Essene and many others confirm the great genetic works of these ancient people.

A New Timing

This doesn't really affect the realism or fantasy of Global warming by Greenhouse gas, but it is another area where evidence has been kept from students to secure some irresponsible course of keeping the status quo no matter how erroneous the information is known to be. What I'm talking about here is nuclear decay as a form of "accurate dating. For some time now, ALL have known the huge issues in this previously established timing baseline. You have been told dinosaurs died 65 million years ago and the beginning of the Mesozoic Era was 300 Million years ago. You were told, over and over and over again. They even proved to you that was truth by telling you lead, potassium, and even carbon isotopes decayed at a set rate; just see how much of an early isotope is left and read the date. Besides, the dinosaurs are buried underground and turned into stone so there had to be a long time for that to happen.

Ancient Earth

While the earth is ancient, it is definitely not as old as has been told to you. Many of geologists today still tell you that radiometric dating has narrowed the age of Earth to about 4.5 billion years, give or take a couple of percent. We now know that the dating method is inaccurate and scientists not pursuing that vain truth I talked about earlier are refining the timing more and more each day. The Earth and everything in it is much younger and so are the characteristic stabilities of the planets in our Solar System. Researchers at Purdue and Stanford have found evidence that **radioactive decay rates are not constant at all.** <u>On December 13, 2006</u>, a magnificent solar flare flung radiation and solar particles toward Earth. Measuring the decay rate of manganese-54 during the flare proved to be very interesting as the decay rate dropped during the time of the radiation fallout. It was determined that solar neutrinos zipped through space and affected Mn-54's decay rates used in the experiment. Just think about this. They were testing a single solar flare event and the change was significant. The sun has these things all the time. It was also found that the decay rates of silicon-32 and radium-226 showed seasonal variation, according to data collected at Brookhaven National Laboratory on Long Island and the Federal Physical and Technical Institute in Germany. This error was just the material sitting there with almost no outside interference. Wood buried in igneous rock in Queensland Australia has been dated to <u>40 thousand years</u>, while the basalt around it dated to <u>45 million years</u>. Both dating subjects should have given the same date, since the igneous rock was formed at the same time the wood was buried. Many of the "data-ologists" don't tell you about major errors like this.

Lava Errors

Excess argon-36 was found in three out of 26 lava flows in recent times. So <u>Argon/argon testing would show a much older date that actually was "KNOWN"</u>. This is believed to be because there was too much of the argon-36 in the first place. In the Grand Canyon lava flow testing showed lower levels of lava were younger than the top layers. At different volcano sites, that had eruption in 1949, 1954 and 1975.

The same thing was noted. Geochron Laboratories of Cambridge, Massachusetts dated these samples. Even though the oldest of these samples are just over sixty-years old, the lab tests provided ages that ranged from 270,000 years to 3.5 million years old. Additionally, we go to Mt. St. Helens and its eruptions in the 1980's. Samples there gave old ages in the range of 300,000 to 2.7 million years. Hopefully, you are beginning to see that we know less about how old we are than you believed before reading this. If neutrinos from a single solar flare can make things look older, what if the entire Earth was closer to the sun? I know that sounds odd, so just keep it in the back of your mind right now as we try to find some standard for dating.

Nuclear Decay a Bad Timing Method

Today we know that the nuclear decay dating of things including Electron Spin Dating and Uranium Dating, Thorium Protactinium Dating, Oxygen Sediment Dating, Lead-lead-lead Dating, and Argon Dating [which we originally used to date the ages of the Earth] are terribly flawed. The old standard carbon 14 dating also seemed in jeopardy. Dating beyond about 30 thousand years was much younger than tested. If there had been nuclear events [bombs or even volcanic eruptions] the apparent timing was changed drastically. Other methods had to be employed to determine how everything should be timed, but classroom information was not changed. That would confuse the students. I'm going to prove to you how you have been lied to. This will give you a better understanding of the lengths some will go to when they believe something, no matter what the evidence shows.

Standard Geological Timeline

Era/Period/Epoch	Time (M yrs. ago)	Time (T yrs. ago)
Archaeozoic Period	5000-1500	50,000-3000
Proterozoic Period	1500-545	3000-1000
Cambrian period	550-500	1000-900
Ordovician period	500-440	900-800
Silurian period	440-410	800-700
Devonian period	410-365	700-600
Carboniferous	365-300	600-500
Permian period	300-250	500-400
Triassic period	250-212	400-300
Jurassic period	212-145	300-200
Cretaceous period	145-65	200-100
Tertiary period	65-1.8	100-40
Pleistocene period	1.8-0.01	40-10
Holocene period	0.01-0	10-00

The middle listing of dates is the "STANDARD" that had been presented in our classrooms, while the last column shows a somewhat closer, more accurate time line that has been verified by non-nuclear decay methods. Even with the mountain of evidence showing how nuclear decay cannot be used, the middle timing is still heralded as the master in many schools and books being used to teach our children without basis. I know it is difficult to believe historians, scientists and teachers would keep these things from you, like how does greenhouse gas affect our planet, so let me tell you a little more.

Stratigraphic Positioning

Besides Nuclear decay, the main way scientists used to determine "age" was by Stratigraphic Positioning. This is the determination of age by position, depth, and material consistency. MANY TIMES this is the only method for cross comparison that was thought to be reasonable for confirmation of Radioactive decay. Scientists simply determine the depth of objects, or features near the object,

28

or number of lava flows, or similar geologic characteristics and use the depth as a time gage. This type of comparison may not have a very high level of accuracy, but seeing things in different layers seem to show when something died. If something is lower, it is older and newer is newer. Added to this method is something called the K-T boundary, where iridium chalk was deposited from an ancient meteor that struck the Yucatan around the time the dinosaurs died. Scientists have been using this for a long time when, all of a sudden, there were trees found that were going the wrong way. The next set of pictures shows some of the unfortunate trees that must have died repeatedly to be deposited perpendicular to all of the stratigraphic lines.

Some try to state the trees simply fossilized while standing for MILLIONS of years as the ground built up around them. [20 points on the BUNK meter!]

Distance to the Sun

If neutrinos from a single solar flare can make things look vastly older, what if the entire Earth was closer to the sun a few hundred thousand years ago? I know that sounds odd, so just keep it in the back of your mind right now. Right now, I'm going to provide you with a more logical way the Pacific Ocean was made at the beginning of the Triassic

Period as our planet rotation was not stable. That is where Ice samples come in.

Ice Core Dating

Although the task is tedious, ice can be examined just like tree rings. Each summer ice changes its consistency. H_2O (16) is more concentrated in the summer while H_2O (18) is more concentrated in the winter. This gives us indication to the level of CO_2 which in turn allows us to understand something about the temperature levels. As the yearly cycle has freezing and thawing, ice consistency varies each day, seasonally, and yearly, depending on Earth axis and other critical elements. Anyway, scientists around the world started boring holes in ice. The most coring is done in Greenland and Antarctica. A sample is shown below.

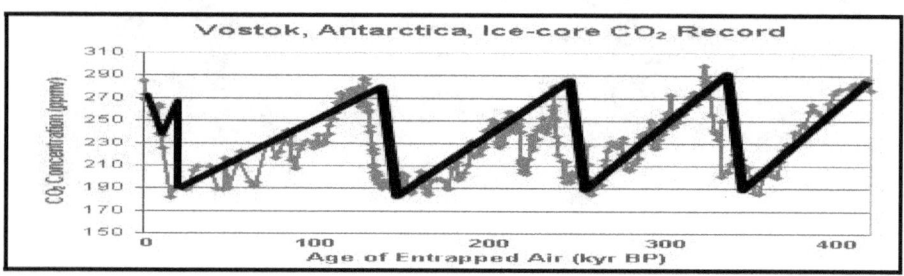

If you look closely you can see that about every 120 years there is a major change in the environment. This was found at both Antarctican and Greenland Ice cores and the dating is by seasonal changes rather than nuclear decay. Bah humb! You say! Well, what if we see confirmation?

Hawaii Hotspot Track Dating

Hawaii is not a tiny group of islands, but instead is an indicator of where the Earth magma has a hotspot. As the crust moves differently than the stuff below, the hotspot relative to the crust moves and each time the hotspot burns through another piece of crust, a volcano erupts which seals off the area after a time and an island is made for a few thousand years. This travelling hotspot known as Hawaii is show next. The descriptions provided shows what was happening along the way. Because the hotspot moves perpendicular to the axis of the Earth we also know how the earth was spinning as shown by the lines in the first graphic below, but the actual timing is not described here. I placed some general times in the second graphic, but let's see if they make sense.

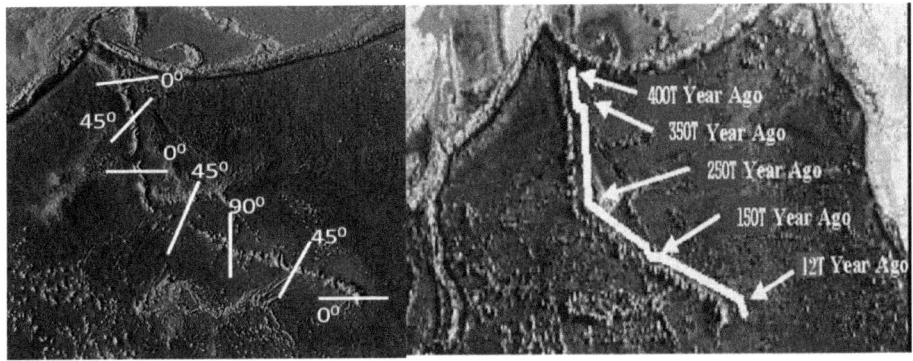

Let's compare the Earth shifts with the Ice core data. Man-oh-man; it seems they match. I think you still believe in nuclear decay so we will look farther.

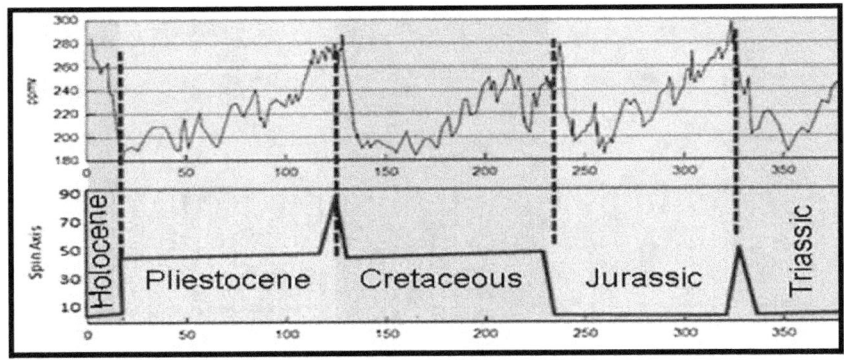

The Atlantic Ocean is getting wider about an inch a year, averaged worldwide. While the building of the great mountains has little to do with the normal tectonic plate "drift" We can pretty accurately measure the widening ocean in various ways including measuring distances between matched magnetic landmarks on either side of a widening gap on the ocean floor. The old theory indicated that 180 million years ago the continent Pangea began splitting apart and has been drifting ever since. In so doing, the landmasses of the Western and Eastern hemispheres separated and opened the Atlantic Ocean basin today.

Plate Tectonics

Plate tectonics tells us the outer hard crust of Earth consists actually of a dozen or so distinct, hard plates that drift individually on hot, deformable rock. An unequal distribution of heat within Earth moves the plates. The boundary between the plates forming the Atlantic Ocean is smack down the middle along the Mid-Atlantic Ridge, shown as the hashed line in the figure below. The ridge is where we must look to find a widening gap, which accounts for the widening ocean.

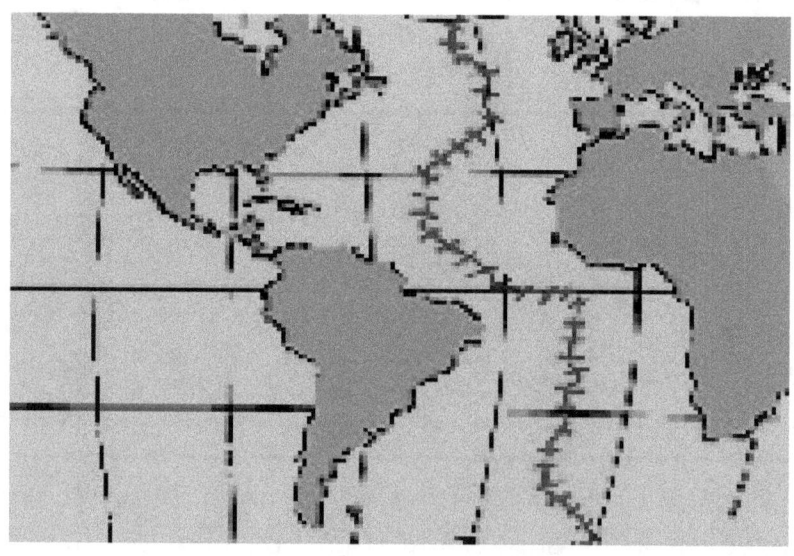

That is where we measure the rate of separation. Where the plates separate, white-hot soft mantle oozes up from great depths within the Earth to fill the gap. The molten rock cools slowly into new slivers of sea floor. This happened over and over again through the eons. That's how the Atlantic Ocean widened-by a spreading sea floor. Iron-rich rock has a peculiar property; heat it above its curie point of 580 degrees Centigrade and it loses its magnetism. When it cools the rock gets re-magnetized in the direction of the existing Earth's magnetic field. So it's a magnet with the poles aligning with the poles of the Earth at the time of the cooling. The neat thing about this is: the magnetic field of the rock, once cooled, stays frozen in this orientation. It becomes a record of the Earth's field at the time of its cooling. The first graph below shows how the magnetic field has changed over time. Certainly we cannot get an actual time, but a relative timing is very good. What if I told you this matched up exactly with the Ice Core and hotspot data?

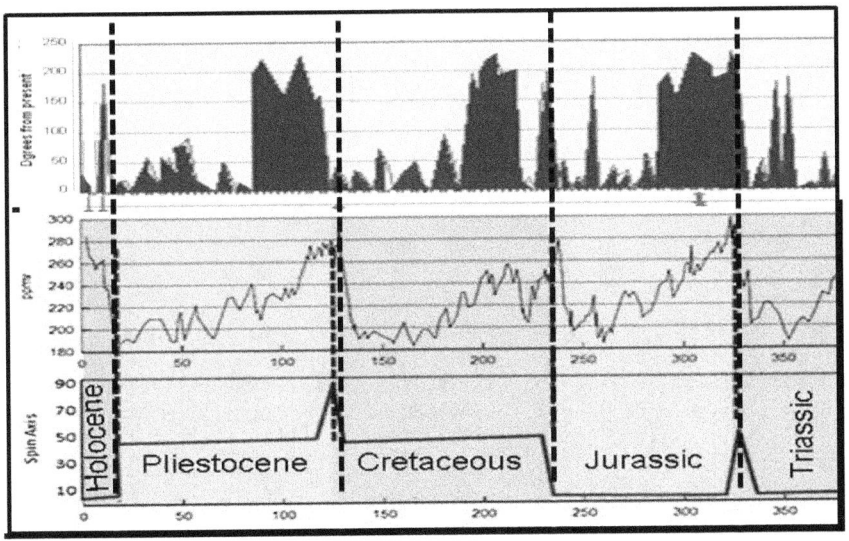

Marine Isotope Stage [MIS] Dating

Some people may still be reluctant to give up what the schools have been preaching so very long, so I thought I would bring out one last attempt at presenting sanity. Large numbers of scientists around the globe are doing Marine Isotope Stage timing by digging in dirt. It seems looking at the levels of Oxygen 18 shows how hot or cold a point is in time while checking relative Oxygen 18 isotopes in Calcite [which just happens to be the main ingredient in seashells], one can tell just how many of the things were here during each period. Checking around the globe has given us a good map about climate and number of seashell, which correlates to number of animals in general so it is easy to see where extinction periods are. Guess where they line up? Time's up! They are an almost exact match as shown below. MIS levels are shown next above the ice core samples, the hotspot data and the magnetic field shift data Massive drops in O_{18} mean massive drops in sea shells and all other life. Notice there is no extinction period between the Tertiary and Pleistocene Ages marked by Cro-Magnon appearing. Please say you see a comparison.

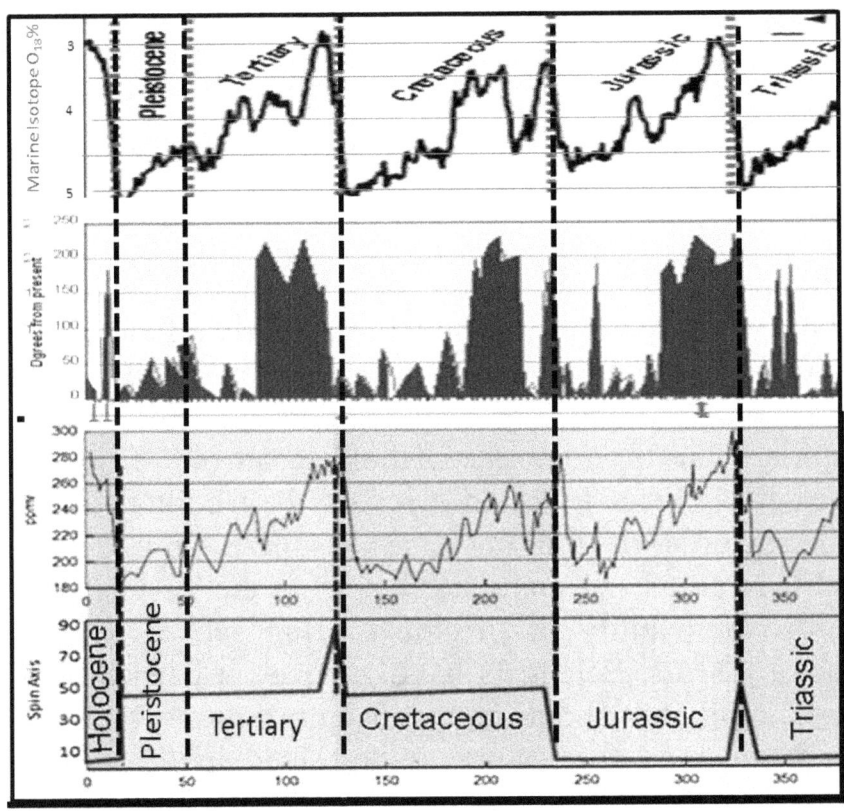

I know these dates are not what you were told and I know the idea of using nuclear decay to date things has made a comfortable geologic record, but the pieces don't fit. The mutation of mankind doesn't fit and new discoveries of non-fossilized dinosaurs don't fit. Another thing that is odd is that there should be an easy distinction in the geologic record to show when 1/3 of the planet was ripped away.

More Ice Core Sampling

I'll bet you are wondering how in the world all these similar timing components could have been known and you were never told!!! If you are beginning to see how scientists, historians and professors will hold to an old "truth" even when it is found to be in error as they built their entire understanding of the universe on these now destroyed truths. Now let me continue this same tactics to help you see the probability of Prestonia being sucked into outer space by Mars. Scientists have gone even deeper into the Ice to find out about more ancient times. They show something very interesting as described in the following graphic. The chart below shows a cyclic changing in the earth's temperature and Carbon dioxide levels over the last 800 thousand years, but notice something STRANGE! What we see is that after 400 thousand years ago, there are very distinct and abrupt thermal changes every 120 thousand years associated with massive extinction periods. The cyclic nature continues before that time, but the events are greatly softened showing the characteristics of the Earth were vastly different before that fateful time. Possibly, this would be something about the larger planet and a smaller portion of the planet core being shifted as the earth spins. Can you see the remains of Prestonia going into space and the Planet Mars being ripped in half?

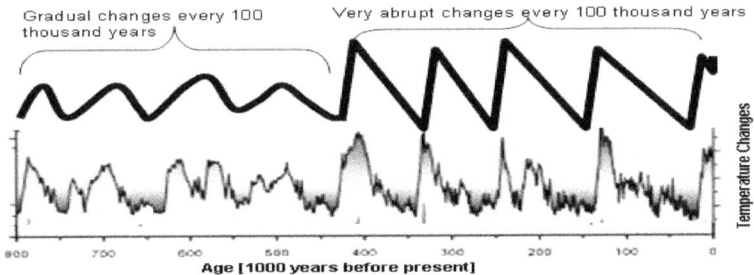

Gradual changes every 100 thousand years

Very abrupt changes every 100 thousand years

Temperature Changes

Age [1000 years before present]

Evidence of a Split Planet

I'm thinking you are skeptical about Mars being split in half. I don't want to get into it much so we can get one with people dying, but here are some images from NASA that may help. The next image is a topographical map of Mars showing that the southern half still has hundreds of thousands of meteor craters and a thick crust while the northern part has been sucked away. Please note that the northern hemisphere is not only smooth, with almost no craters showing this DID NOT happen in the too distant past, but it also is "sunken-in" much worse than our Pacific Ocean. It has a mean surface height 6 thousand meters lower than the mean of the southern hemisphere. Where do you suppose the northern half of the Planet went? The drawing to the right shows how Mars is trying to repair the damage just like our Pacific Ocean is getting smaller.

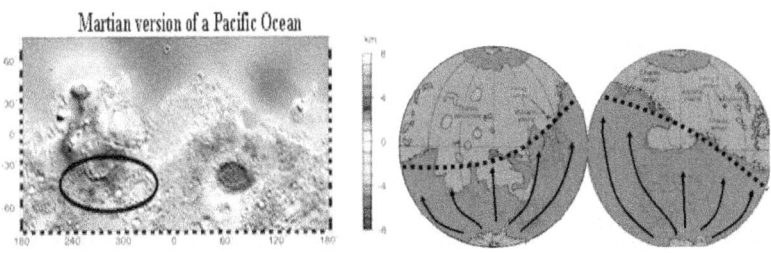

Martian version of a Pacific Ocean

37

Plate Tectonic Issue

As I mentioned, someone once told you the entire Pacific Ocean was made by Plate Tectonics, a plate moved fast and made the mountain range that went from Antarctica, through South and North America and around to Russia and down into the Orient. Ha! Ha! I used to think that made sense because they kept telling me it did, but Plates cannot go in three directions at the same time from a massive explosion and the energy required to push this supposed massive plate in all these directions would have had to be so huge, the Earth would not have survived. The image below shows where this mountain range [supposedly pushed up by plates getting on top of other plates.] Now just for kicks, make this mountain range MUCH bigger as it would have been made before Pangea began separating and making the Atlantic Ocean.

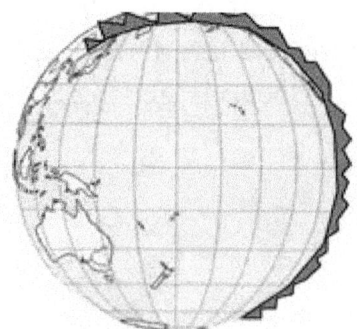

Unstable Orbits-We now know that there is a 100 thousand year cycle that seems to be self-generated by the Earth itself, sometimes the changing characteristics that we can use as time-marks. Every 100 thousand years or so, the Earth gets terrible cold, which sometimes shifts the rotational axis and causes extinctions and or problems. I \'ll show you that these changes and extinctions are done outside the emission of hydrocarbons and greenhouse gas. They are insignificant. We'll look at some of the data in just a minute, but right now just understand that about 400 thousand years ago, the cyclic nature of the Earth cycles changed dramatically. Mathematical models describe this change as having extraterrestrial connections. I know that sounds like little green men, but that isn't what I'm talking about. I'm talking about planets getting to close to one another. The following image shows the first of three of the major ones believed to have been associated with major events most likely involving Mars. To the right shows a second one that happened just before the end of the Pleistocene. While many have pushed these events back hundreds of millions of years ago, from the timing corrections presented in this work, we can assume that the stability of the solar system is only recently been "standardized" into almost circular orbits around the Sun. It is the variable intensity of the sun that has caused much of the confusion in timing, so let's peel back the history just a little.

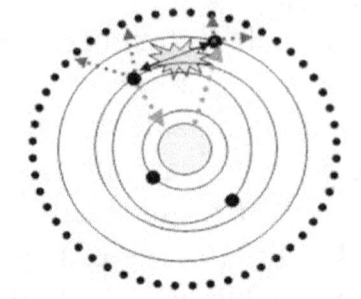 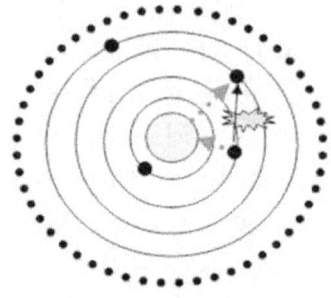

400 thousand years ago Mars almost hits earth-both planets have massive damage- Mars orbit pushed out, Earth orbit come closer to sun- Moon and Asteroids formed

12,000 YEARS AGO- close encounter Venus and Earth. Venus moved closer to sun, Earth farther away. Venus moon destroyed

Four hundred thousand years ago the orbits of the planets weren't circular and on rare occasions, the planets came close to each other. The sun was more intense during this time period and pushed out many neutrinos into the earth's atmosphere making more and more nuclear isotopes rather than allowing them to "decay" as many had previously believed. Hundreds of thousands of years appeared to be hundreds of millions of years because of this massive illusion. The earth during this time was filled with 2 great continents. The first will be known as Pangea, but the second one, on the opposite side of the earth had no name, so I named it Prestonia, at least while we address this unknown mass in this book. Without the second continent, counter balance, the earth would not have been stable and would have wobbled violently. With all the close collisions going on, the Pacific Ocean was made. Mars got too close and pulled Prestonia right out of the earth and then Venus had its turn later.

Mars Mess

From the preceding diagram, notice the 4th planet which should be named Mars-plus as it is much larger than today's

40

planet, comes near the earth that is larger as well. As it passes, the Erath crust is pulled upward into the sky creating the Himalayan Mountain Ridge. We can believe there was massive destruction as the earth environment was rocked by the intrusion. This event triggered the extinction between the Triassic and Jurassic periods on earth and millions of animals were destroyed. I like to think the animals on Prestonia were more evolved than those on Pangea, but we may never know. One thing we believe about this time period is that humans lived and walked on the land. Not only were these people living on the earth, they were civilized to some level and they were giants according to almost every ancient text known as such they certainly used and emitted those deadly hydrocarbons that are brought up all the time.

Pulled up Mountains

On Earth we can still examine the effect of the close flybys of Mars in the form of extremely long mountain ranges. Early theories that mountains were pushed up by plates moving together did not match the positions of the mountains and people began to wonder why the mountains all fell in straight-line patterns. One path is straight along the equatorial line and another goes along the side of the Americas and along the same path on the other side of the world. The figure on the following page shows the 2 great mountain ranges. Plate tectonic models should not go around more than 50 percent of the globe, because there would be no way to push the block, but the tectonic theorists had to make them that way to support the mountain ranges until sanity finally won out.

New math models were able to capture the events that caused something quite different that mountains being pushed up. Instead, the mountain ranges had to have been **pulled up**. A large planetary object strafing the planet made each of the extended mountain ranges. Once the Earth was strafed with the Earth rotating on an axis through the middle of the Pacific Ocean and Asia and a second time when the rotational axis was similar to our present rotational feature. Yes, I did say the earth's axis changed. In fact, we will see that it has happened more than once. The following picture shows the projected path of the Mars close encounter on two separate fly-bys. The wide lines represent the long strings of mountains along the 2 paths. One going from the southern tip of South America through the tip of Alaska and down along the coast of the Far Eastern countries. The second, more severe uplift included the region from the Middle East through Pakistan, Tibet and China. By the time the uplifting got passed China, the uplifting lifted out the section of the Earth that made the Pacific.

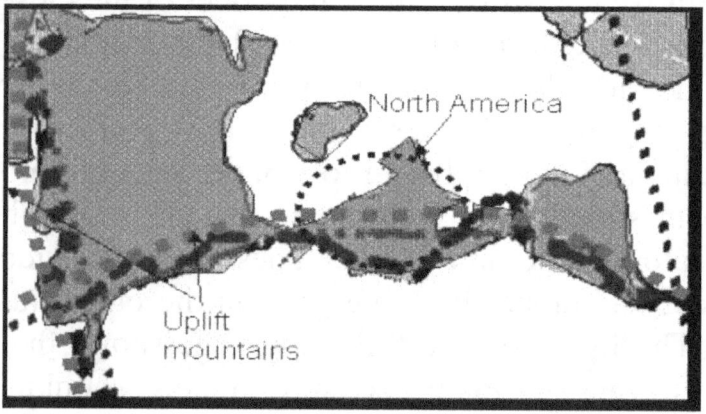

Walking With Dinosaurs

This section tells of another "vain truth" taught that presents the first "modern Man" as Cro-Magnon who entered the scene about 40 thousand years ago. They know there is strong evidence of the previous people, but it doesn't "FIT" in their version of the truth. Rock strata from Triassic, Jurassic, and Cretaceous periods contain literally billions of dinosaur tracks, and actually outnumber bones by orders of magnitude. After all, dinosaurs only made one skeleton and many footprints in its lifetime, so we can get a better understanding of dating from the footprints.

Tracks Found Everywhere-Dinosaur tracks have been found in over 1000 locations throughout the world, on every continent except Antarctica. In the U.S., they are especially abundant in Texas, Colorado, Utah, Arizona, New Mexico, Connecticut, Massachusetts, and New Jersey. It is believed that in the western U.S. alone new sites are being reported at the rate of about 50 per year. Most of these tracks have been found where there once were shorelines large expanses of moist sediment were so important in building proper fossilized tracks. Here is the weird part. Human footprints are being found with the dinosaur ones. Many times these footprints show humans that were huge lived with the huge dinosaurs.

- *At Rocky Hill, Connecticut can be found a track floor that is covered with hundreds of theropod tracks.*

- *In Amherst, Massachusetts* one can find thousands of lower Jurassic dinosaur tracks from the Connecticut Valley of New England.

- *At Glen Rose*, Texas one can see many large Cretaceous carnosaur and sauropod tracks still in their original positions.

- *At Seneca, New Mexico* the site contains hundreds of ornithopod and theropod tracks.

- *Tuba City Arizona* has a site containing many lower Jurassic theropod tracks.

- *In Denver Colorado* several Cretaceous dinosaur trackways can be seen still in their original position.

- *At Alberta, Canada* a vast collection of dinosaur tracks from the Peace River of British Columbia can be seen along with one of the largest exhibits of dinosaur skeletons.

- *In Price, Utah* one can see displays that include about 50 Cretaceous dinosaur tracks collected from coal mine roofs.

On and on we could go, but the thing that is unusual is that many sites have human footprints mixed in. The graphic below shows some of the trackways that have uncovered the evidence of people walking on the same beaches as dinosaurs..

Turkey-In the late 1950's during road construction in Homs southeast Turkey, many tombs of Giants were unearthed. These tombs were 4 meters long, and when entered in 2 cases the human thigh bones were measured to be over 47 inches in length. It was calculated that the person [or Titan] who owned this Femur probably stood at **fourteen to sixteen feet tall.**

Mexico 1925-According to the Washington Post, June 22, 1925, and the New York Herald-Tribune, June 21, 1925, a mining party found skeletons measuring 10 to 12 feet, with feet 18 to 20 inches long, near Sisoguiche, Mexico. These also sound like Titans.

In South Africa, a giant footprint of a woman measuring over 4 feet long has been dated to be from before or during the Triassic period. Pointing to the probability of this being a female human-like species' foot, proportionally the two-legged being would need to be <u>some 30 feet tall!</u>

Chinese Titan Footprints-In Shenmu County in China's Shaanxi Province, in 1967, a man surnamed Qiao went to quarry some stones and found huge human-like footprints encased in the stone heading to the edge of a cliff. Each of the footprints is about 16" long indicating the human was about 9 feet high. Considering the footprints are embedded in the stone, they were determined to be Cretaceous.

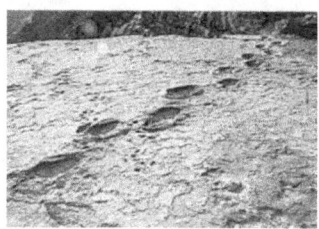 

Giant Bones In Ecuador- A find from the estate of a Catholic priest in Ecuador points to the existence of Titans in Ecuador. A collection of giant fossilized bones was found in the estate of the late Father Carlos Vaca. Anatomists identified one of them as the occipital section of a human skull. They believe that another bone may be part of a massive human heel. Judging by its size, it would have belonged to a giant human that may have been over 7 meters tall [over 20 feet tall]. The bones have been dated to be Cretaceous.

At Inverell, Australia-we find an example of the Titan evidence long, long ago. The footprint below right shows the tremendous size that some of these people got. This particular one was found along with many others.

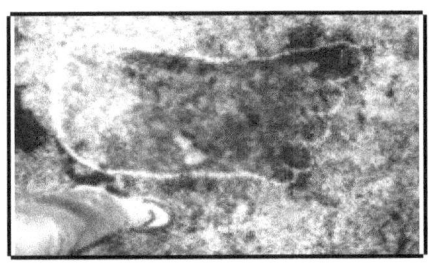 

Australian-In September 1993, another giant-sized human fossilized footprint was added [above left]. Here, embedded in the rock, was a large footprint impression. The track was that of a right foot, probably distorted in the original soft mud, and was 44 cm in length x 30 cm across the toes. There were signs that other tracks had been embedded nearby, but these had weathered away. The imprint was preserved by a lava flow that was "reportedly" dated about the time of the Homo-Erectus. The monstrous human whose single footprint still survives was about <u>11 feet in height</u>.

Giant "Human" Shoes in Australia- Don't believe these people only walked on the beach barefooted. The next photos are from NSW, Australia of shoeprints. Next to the shoeprints are images of the photographers "tiny" shoes. As shown to the right is the general placement of the 2 complete shoe prints and a number of partials that were found. Now, when I found these prints, there were a number of prints and half-prints, which appears to be three sets of different prints, interlocking with each other.

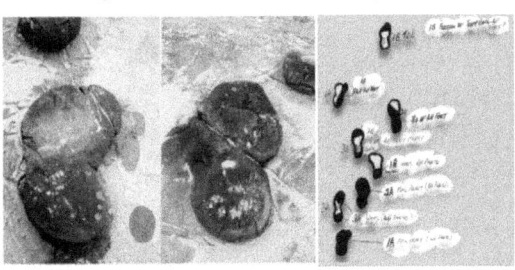 

Another group near this same location also appear to be human shoeprint, but this time, the raised edges show that he was getting stuck in the mud as he walked possibly from the Cretaceous time as well.

Skeletons

Arizona-In 1923, Mr. Samuel Hubbard discovered the remains of giants in the Grand Canyon of Arizona. The discovery consisted of the following: Petrified bodies of two human beings about <u>18 and 15 feet in height</u> respectively. One of these was buried under a recent rock fall which required several days' work to remove. The other, of which Mr. Hubbard took photographs, was in a crevice and inaccessible. The bodies were formed from limestone petrifaction and embedded in sandstone during a very ancient time.

Nevada-In July, 1877, four prospectors were looking for gold and silver outcroppings in a desolate, hilly area near the head of Spring Valley, not far from Eureka, Nevada. One of the men spotted something peculiar projecting from a high ledge. The prospector was surprised to find a human leg-bone and knee cap sticking out of solid rock. He and his companions dislodged the oddity with picks. Realizing they had a most unusual find, the men brought it into Eureka, where it was placed on display. The stone in which the bones were embedded was a hard, dark red quartzite, and the bones themselves were almost black with carbonization showing its great age. When the surrounding stone was carefully chipped away, the specimen was found to be composed of a leg bone broken off four inches above the knee, the knee cap and joint, the lower leg bones, and the complete bones of the foot. Several medical doctors examined the remains, and indicated that they had indeed once belonged to a human being, and a very modern-looking one. But for us the best part was their size: From knee to heel, they measured 39 inches. Their owner in life had thus stood <u>over 12 feet tall.</u> Compounding the mystery further was the fact that the rock in which the bones were found **dated to the era of the dinosaurs**, the Cretaceous or earlier. The local papers ran several stories on the marvelous find, and two museums sent investigators to see if any more of the skeleton could be located. Unfortunately, nothing else but the leg and foot existed in the rock. Again, this is just a tiny sampling of huge amounts of information.

Ancient tests tell us most about what we know of this ancient group. We understand that they had great cities and a high level of civilization. They also had scientists and modified animals making some of the largest and most odd looking "mistakes" ever seen. A number of writings tell us they became spirits at some time and the timing corresponds to the entry of the humanoid beings called Watchers. While some became these watchers, some still remained as Titans until the dreadful war that happened about 120 thousand years ago. According to the book of Isaiah and Genesis, the entire world became desolate and all the cities were completely flattened. Many ancient texts call this time, "*The War between the Giants and the Gods*". Titans had been the undisputed rulers of the land for hundreds of thousands of years and all was gone. While almost all the physical evidence of the civilization of Titans was lost with time, occasionally things do pop up. Here are a few.

CO2 Generating Civilization

These ancient giants were not lumbering fools as some of the Greek histories try to portray. They were very civilized and skilled in all types of science and technology. As such they produced thousands of tons of those nasty hydrocarbons that are being talked about today. Here is a sampling of some of what has been dug up showing their advanced levels on manufacture and science.

Building Materials and Art

Iowa-1897-A large stone [2x2x1feet] with multiple faces of an old man carved on it and a grid pattern on the remaining area was found 130 feet down in a coalmine. The estimated age was Cretaceous. [Below left]

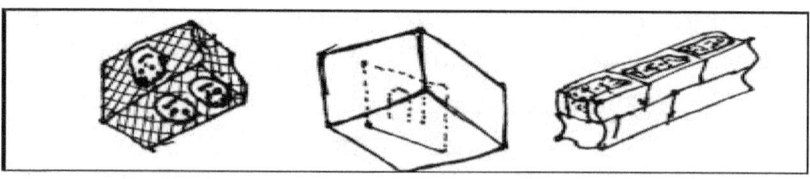

Philadelphia 1829-A 30 cubic foot piece of marble was excavated from a depth of 60 feet. Inside the marble was a straight edged rectangular indentation. After a section of the marble was carefully removed, it was found that 2 distinct heavily engraved letters similar to an "I" and a "U" eleven inches long and 5.8 inches deep were on a square base. The estimated age was Cretaceous.

Oklahoma-1928- [Above Right] A block wall was found almost 2 miles deep in a coal mine. Each block was 12 x 12 x 12 inches polished on the outside and filled with gravel on the inside- There were multiple reports over 150 yard length of the same wall. The estimated age of the wall was Cretaceous.

Strange Geode

The old battery in a geode-The picture next is some type of power conversion device found **inside** a geode, in California. Below the geode is a drawing of x-rays of the geode showing the elemental parts. These include a spring, core, plate, and electrical insulator. The same parts as you would expect in a battery. Maybe this is a new way to package batteries, but it takes a long time to complete the package. Both of the objects are extremely ancient and certainly, before we originally thought that everyone used electricity. The central metal core surrounded by the white material looks like a battery. Whatever it was, it was electrical. On the right is a drawing of the parts and a size comparison to a standard D-cell battery.

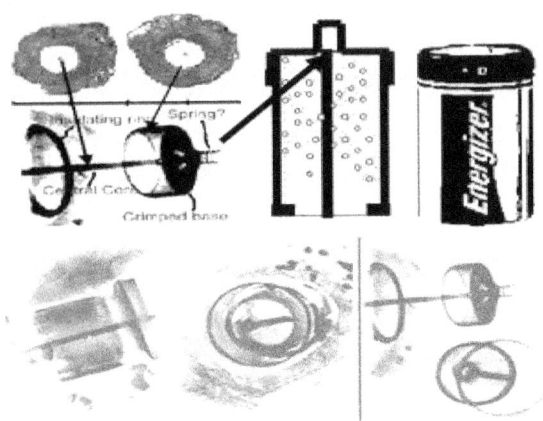

Ancient Nuclear Plant-The last thing I want to bring up is the massive nuclear plant in Oklo Africa. A nuclear plant with at least 16 different areas of uranium depletion has been found and this massive thing is dated to before the end of the Jurassic Period. The plant complex is unusual in that there is no nuclear fallout. I know you were thinking these people were so backward, they had to rely on simple batteries for power, but I'm talking about nuclear plants, bombs, nasty wars etc. While built during this time, these plants were, evidently used over the years by others. In Oklo, Gabon, Africa, we found 16 depleted uranium pockets or processing areas inside some caves. Very quickly, scientists backpedaled and came up with the story that this was a naturally occurring nuclear plant, just like any other natural nuclear plant. Wait a minute!!! There aren't any! These nuclear reactors were estimated to have produced on the order 0f 1,000 megawatts, comparable to a large modern plant. All this could have been used to light houses or for war. The normal Uranium 238 was "processed" to Plutonium and "enriched" to Uranium 235 allowing energy to be provided and used [by someone]. The following graphic shows where the processing took place. I know it just looks like blobs, but believe, me this was put to use in the olden days and we have proof to look at later as it would not just be used by the Titans, but also the group called the Anakim and finally it will be used in the most devastating war of the modern Age as man tried to destroy his home.

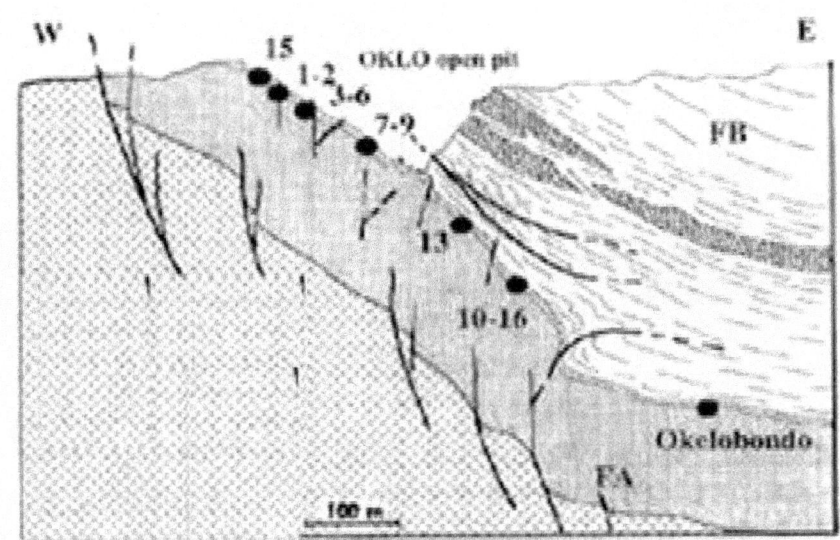

Death Before Chalk

Most know about this iridium chalk layer that is found at the stratographic position determined to be the end of the Cretaceous. According to many, this comet from outer space hit the Yucatan and blasted a hole in the Earth. It hit so hard that the comet exploded and send iridium all the way around the earth marking the end of the dinosaurs. Some of that seems reasonable. Whatever happened, the Iridium in the boundary layer can also been attributed to another source besides a meteor. The earth's core could have done it if the core somehow got to the surface. That is where India comes in. The entire country of India wasn't here before the K-T layer was formed. Whatever hit the earth near the Yucatan caused the earth to split open and start spewing out magma. This continued until a newly developed landmass was formed that we call India. The air was filled with soot. Today, hundreds of cubic miles of magma <u>still fill the area called the Deccan Traps</u>. [Yes, I said MILES] By the way, dinosaurs weren't all killed by the Chalk as some have indicated. Many, many dinosaur fossils have been found <u>below</u> the K-T layer showing they were killed a thousand years or so before the Chalk event, DURING some massive nuclear war as ***more and more hydrocarbons were expelled into the atmosphere.***

One thing that is known by scientists; while Iridium is only found on meteors and deep inside the earth, the amount of iridium found around the world in the K-T layer is far too dense to have come from a single meteor. The reason they don't tell you is that most of the iridium came from the Earth and the Country of India tells the awful story. The graphic following shows how chalk killed the giants, sent massive amounts of hydrocarbon blasts into the Ozone, and eventually created the country of India.

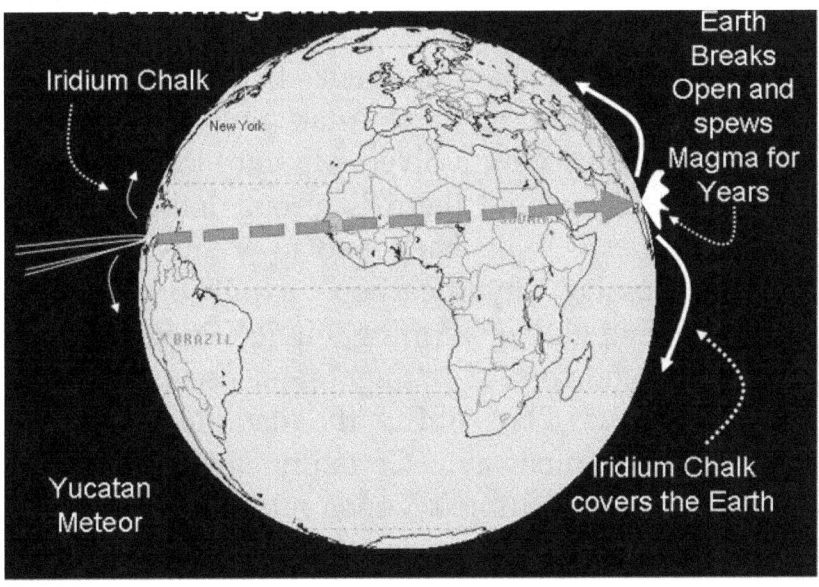

When the Yucatan "Meteor" hit, the other side of the world belched forth millions and millions of tons of Magma from a massive opening. This huge pile of magma is called the Deccan Flats. Originally, this massive pile of "earth insides" covered over 2,000,000 square meters and contained about a million cubic kilometers of magma rich in iridium. This sort of thing has happened before man was on the Earth, but this time was the only one since mankind was created and it seems to have caused the destruction of the Titan people if any remained after the Great War period that ended what the Greeks and PreMaya called the Golden Age. The Titan People were all gone, but the Earth remained solid.

A New Group of Rulers

As the Titans disappeared, the ANAK or Anakim People came along to control the world. Certainly, there are many other sub-races. These people were almost as large as the Titans and they lived during the Tertiary and Pleistocene Age. Many died at the end of the Pleistocene as another great extinction showed up in the Ice core sampling, but before we get to that, I want to examine another hydrocarbon explosion as these Anak guys initiated a huge war and Venus even got involved. Up until this time, Venus was a livable planet and there is strong indications some people lived there. What happened to Venus was not the result of massive hydrocarbon expulsion, green-gas phenomenon, and global warming, but it helped those trying to describe the horrors of letting Freon escape or burning coal so historians CHANGED the data to show Venus dying as thermal runaway. Certainly, there was massive thermal runaway that turned Venus into the melting Planet it is today, but you need to understand more of the story.

Radioactive Dinosaurs 120 Thousand Years Ago

While I don't want to get into the details of the ancient war that ended the reign of the dinosaurs, let me just give you a little information by talking about radioactive dinosaurs. First, some try to say as bones are mineralized, they suck up radioactive materials and become more radioactive than the ground so people go around finding dinosaurs with a Geiger counter. Many Cretaceous Period dinosaur bones are radioactive, but most areas where they are found are not. Go figure! During the Mesozoic Period of the dinosaurs,

56

scientist readily admit that the Oklo nuclear plant was running, but no one wants to say is if animals were radiated, nuclear products must have been in use. At the same time, the Bible indicates the war that killed most of the dinosaurs, and left the entire world without form must have been a hum-dinger as it made many of the dead animals radioactive.

Isaiah 14:16-17-Is this the man that made the Earth to tremble, that did shake kingdoms. That made the world as a wilderness, and destroyed the cities.

Jeremiah 4:23-27-[near the end of the wars] I beheld the Earth, and, lo, it was without form, and void; and the heavens, and they had no light. I beheld, and, lo, there was no man left, and all the cities thereof were broken down

Nag Hammadi-The heaven and Earth were destroyed by the troublemaker that was below them all. The sixth heaven shook violently- when Pistis [God] knew about the breakage, she sent forth her breath and bound him [Satan] and cast him down into Tartaros.

Strangely this horrible War did not send the Earth into thermal runaway.

Another War

Below is a couple of the "retiming-of-the-earth" graphs used today. The one on the top is from Paleo-magnetic readings from the middle of the Atlantic Ocean that do not rely of nuclear decay and the second is one of many Ice Core samples from Antarctica which characterizes time by seasonal ice melts. They show important changes in the atmosphere that could have mutated DNA of all living organisms and caused massive extinctions.

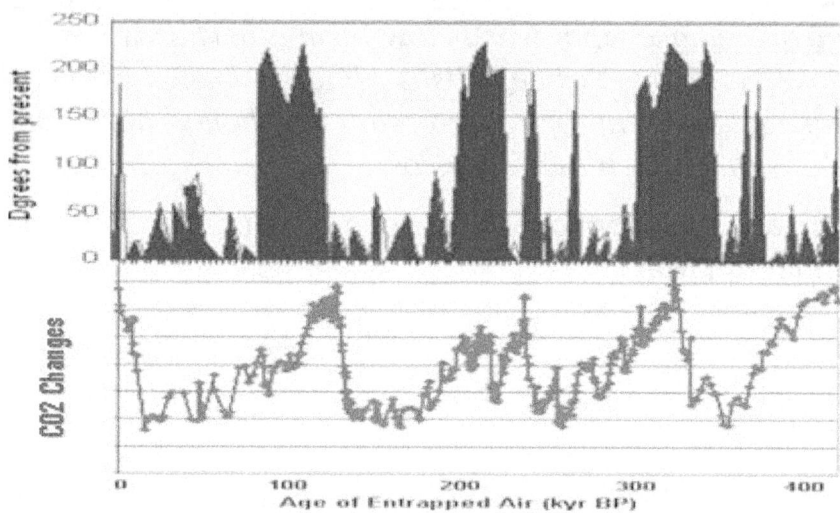

From the cyclic charts, we can easily see the following major disruption periods that would have allowed more mutation:

- End of Triassic[330K Years ago]

- End of Jurassic[220K Years ago]

- End of Cretaceous[120K Years ago]

- The Destruction of Venus[11K Year ago]

- Worldwide flood/End of Pleistocene [10K Years ago]

The last one to recognize was the very last shift in the Earth's rotational axis. It seemed to have happened a very short time after another massive war that ended as Venus [the Planet Rehab in the Biblical history] was destroyed. Another massive war happened 6 thousand years ago. I know you have not been told about these wars, but the evidence is much more massive than presented in this overview what we see is the earth has violent shifts. As the atmosphere settles from this shifted rotation, cosmic rays are allowed to enter the atmosphere. Cosmic rays and cells simply don't work well together. Many times we see almost instantaneous mutation of DNA during this type of intrusion. Besides cosmic waves, we can understand something else. Every time massive wars occur and the Oklo Nuclear plant was used and nuclear fallout not only mutated cells, it also established massive hydrocarbon emissions wherever bombs were detonated. I know all this ancient war stuff seems bizarre to you, but there is ample proof of the events----- even more that the so called complete details of our upcoming destruction by global warming.

Greenland Confirmation

To help confirm the more recent events we need to look at the temperature curves from Ice cores in Greenland. This shows the massive change 11 thousand years ago as the Earth Shifted. A second change is evident around 10000 years ago. Around 6 thousand years ago, another small dip in temperature shows the tell-tell signs of a massive war. I know it doesn't look like much here, but let's look more closely.

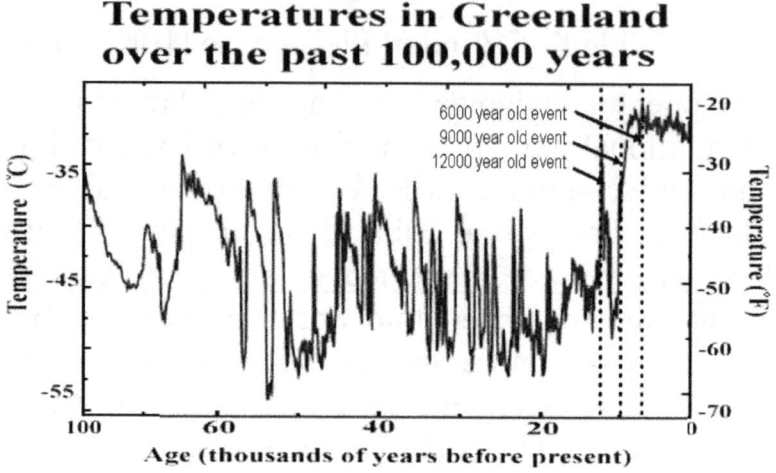

Respiration

The last charts go back to Antarctica to look a respiration just a little. In particular, we need to look at the green gasses CO_2 and CH4 closely. The following multiple year tests of these gases in ice core sample. Notice the huge dip in respiration around 6 thousand years ago. The book of Jasher indicated that 1/3 of the population of the world died in this massive war. Tree-growth samples and similar tests show that during this time there was little or no growth. If fewer animals were producing these gases, there was substantial devastation not clearly shown in the thermal maps, but "something horrible happened just the same.

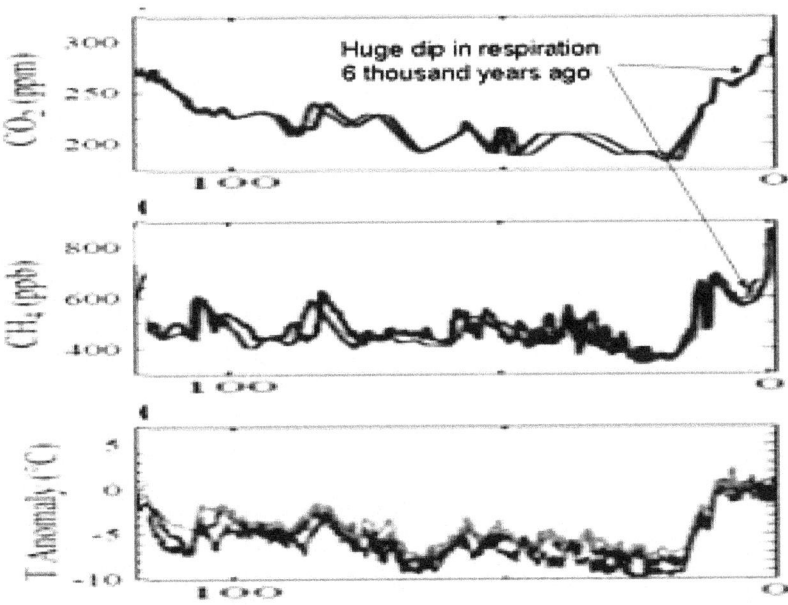

Methane [CH4] reductions not only show reduced numbers of people and animals, but it also shows that the civilizations became less industrialized as populations were thrust back into a Stone Age existence for a time. Most anthropologists are not even looking at this type of data to understand what is happening with our DNA and no thermal destruction alarmists are looking at this data even with it right in front of them.

- *Why would they show there was a drastic reduction in world population 6 thousand years ago when no thermal increase was noted?*

- *Why would they bring up these wars and shifts in the earth's axis when this would give alternatives to data they used to establish dooms day?*

Earth Shift

As we think about the Earth thermal increases and decreases, we need to not only look at Cow and Dinosaur flatulence; we need to look at how the earth axis shifts. For this, we need to look at the Hawaiian Hotspot and understand that the trail of this important hotspot goes perpendicular to the rotation of the Earth. The hot spot I'm talking about here is the one that has made ALL of the Hawaiian Islands and Midway Island. The inside [liquid] of the Earth is rotating slightly differently than the outside. The hot spot itself is part of the internal core of the earth which is moving at this slightly different rate. While this adds a certain stability; for us, it provides and a continuous trail as the hotspot

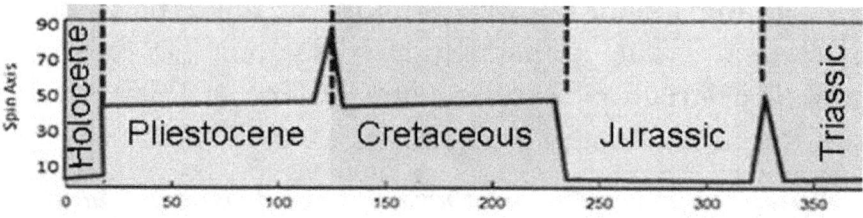

Notice that for a short time [a few thousand years] about 100 thousand years ago. Antarctica was probably warm between the Tertiary [just called the Pleistocene in the chart above] and Cretaceous Period. Sure enough, animals from that time have been found under the ice---just sitting there waiting to be found. Looking at the Ice receding during this time would be stupid as a substantial amount of the ice went away. The graphic below tries to show some possible major earth "settling" points and general information about those spin axes. For instance, notice that the earth spin goes along the east coast of the United States 12 thousand years ago and shifts to where it is today very shortly after than time. This will be important later as we piece all of this together and try to interpret critical timing to help us reevaluate the haplogroup expansions and the DNA mutations that occurred to make us who we are.

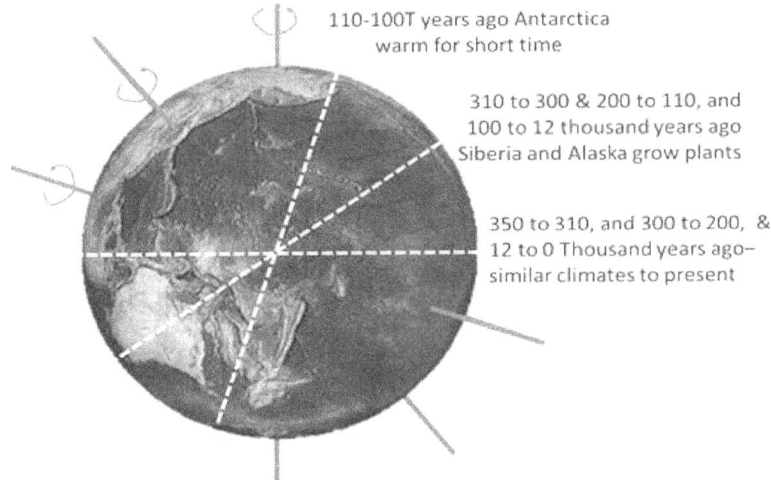

110-100T years ago Antarctica warm for short time

310 to 300 & 200 to 110, and 100 to 12 thousand years ago Siberia and Alaska grow plants

350 to 310, and 300 to 200, & 12 to 0 Thousand years ago— similar climates to present

Additional Information-The other thing we see is that the quick frozen Mammoths and the location of that Garden of Eden make sense.

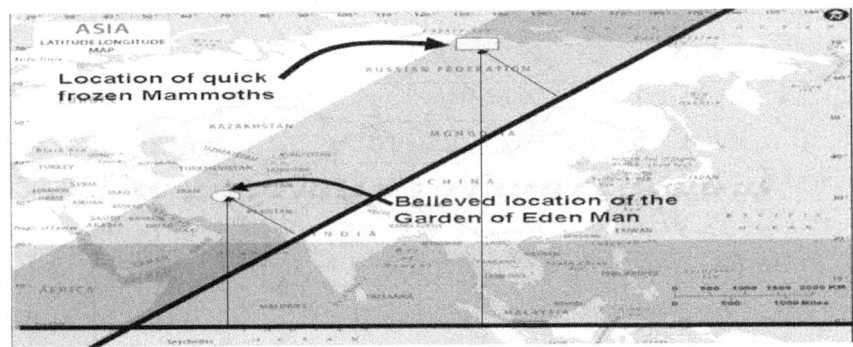

Garden of Eden

From written texts we can determine the Garden was located somewhere between Iran, Pakistan, and Afghanistan. Notice that a location between Iran and Afghanistan had near perfect location for building a beautiful garden 40 thousand years ago [until about 10 thousand years ago]. The slanted dark line indicates the Equator while the lighter area parallel to it shows the region of comfort. But the shift registered by the Hawaiian hotspot shows how it was destroyed 10 thousand years ago at the end of the Pleistocene Age. At that time, the rotation turned to where it is today as shown by the straight equatorial line and our new "comfort area". The garden would have been demolished overnight.

Mammoths

Notice that mammoths chewing on the grassland of Siberia were quickly thrust into the Arctic as the earth axis shifted. While the Mammoths were now in the Arctic, Europe came closer to the equator so the Ice melted quickly. Places like Egypt had very little change in distance from the Equator during this horrible event.

Sorry This Part Sounds Religious

At the beginning of the Pleistocene and new human came into being "miraculously". We call him Cro-Magnon, but the Bible called him Adam. This was said to be the 8th period after a horrible war according to the second chapter of Genesis. More wars continued during the Pleistocene and many died. *This would have caused huge stress on the environment and one would thing global warming would destroy everything.* The Bible indicated that the planet Rahab [Venus] was destroyed and it may have helped the eventual end of the Pleistocene as hundreds of thousands of meteoric chunks hit the Earth and much of the land was on fire. Dozens of historic and religious records describe the war and the end of Venus. Over 500 thousand craters called the Carolina Bays show the onslaught and massive mutation, massive fires, and huge plumes of hydrocarbons to destroy the Earth, but it kept its cool.

Rebuilding

10 thousand years ago data indicates the earth axis shifted and water erupted around the world from the air, the reshaping of the poles and the disruption of a ½ million meteors. Mile high tidal waves killed almost everything. At the end of the Pleistocene, there were boats being made to weather the massive floods that were expected. Noah and his Cro-Magnon family would be saved along with others around the world as the Earth shifted, Mammoths were quick-frozen. Sumerian history indicated the Anunnaki people huddled together in fear in flying machines as the Earth was flooded. They cried out, but there was only rain. Massive tidal waves reached the tallest mountains. We are told it rained for over a month before some could reclaim the land. Noah floated for another 11 months before finding land suitable for him to make a vineyard and raise 3 of his children. With all that, the earth STILL did not go into a global thermal runaway.

Un-Fossilized Dinosaurs 20 Thousand Years Ago

That brings us to "fresh" dinosaur skin and tissue. Some try to connect the 2 and says dinosaurs didn't die until 20 thousand years ago. Bah Humbug! [Sorry for the outburst!] That being said some dinosaurs did not die at the end of the Cretaceous or they would all be fossilized. As just mentioned there was another war just before the end of the

Pleistocene, 10 thousand years ago, which places their destruction during the time of the preflood Cro-Magnon people. A large number of accounts describe the wars and we still have the Oklo plant sitting out there making weapon usable nuclear products. It's not hard to put 2 and 2 together and see that just like was done before, dinosaurs were remade during the war. I don't know what they used them for, but we are starting to find these animals all over, so don't discount the obvious. The huge, unfossilized dinosaurs being found today are radioactive as they must have died during this horrible war. If you haven't been told about these newer dinosaurs, as of April 2014, researchers have begun finding flexible and transparent blood vessels, red blood cells, many various proteins including the microtubule building block tubulin, collagen, the cytoskeleton component actin, and hemoglobin, bone maintenance osteocyte cells, and powerful evidence for DNA in unfossilized dinosaur remains. Blood vessels from a T-Rex are shown next.

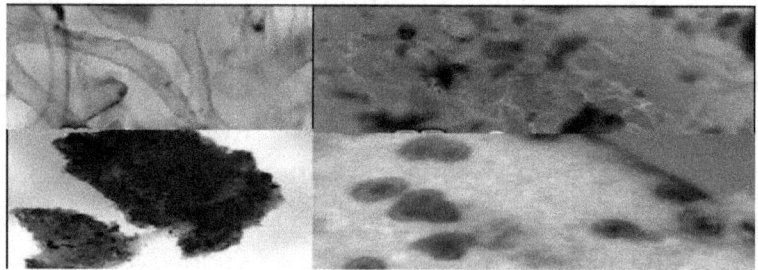

The list of dinosaurs that were, apparently remade during the Pleistocene keep growing and now include; Hadrosaur, titanosaurs, ornithomimosaur, mosasaur, triceratops, Lufengosaurs, T. Rex, and Archaeopteryx. Dinosaurs were getting radioactive, cities were being destroyed, hydrocarbons were expelled by the truckloads and then there was space.

More on the War

Zadspram- Iranian Bible*-And God said, "I will smite thee, Satan, **and the creatures** which thou thinkest have produced fame for thee. <u>I will destroy everything about them</u>" And Satan having darkness with himself, he brought it into the sky and left the sky so to gloom that the internal deficiency in the sky extended as much as <u>one-third</u> <u>over the star station</u>. And Satan was confounded and he fell back to the gloom. Time made the creatures of God moving, distinct from the motion of Satan's Creatures. After **<u>the noxious creatures</u>** <u>died, and the poison there form was mixed upon the Earth and as Satan came thirdly to the Earth which arrayed the whole Earth against him</u>—since there was an animation of the <u>Earth through the shattering</u>.*

The Zadspram expands the Genesis story. Satan sides with 1/3 of his troops in the Star Station, he could attack the Earth more easily. Notice the reference to the 1/3 extension of the star station. If we put that with the other twenty of so descriptions of this war being fought, even on stations in outer space or at nearby planets, we might get a better feel about what happened to Venus and how bad this war was.

Venus the Greenhouse Lie

I know you have heard this one. If you burn too many trees or use too much coal the earth will become like Venus with something we will call the Greenhous Effect--- greenhouse gases increase the heat which produces greenhouse gases, etc. etc. etc. Lie! Lie! Lie! The problem is that there is a tiny bit of truth so it sounds right and good only Venus is sitting there to be a reminder. Once lush and tropical, it is now turned on its side, almost has its rotation stopped, a massive gash almost splitting it in two, and 700 degree temperatures that have boiled away the atmosphere, water, and anything else. <u>First let me say greenhouse gas cannot halt the rotation, turn it sideways or split the planet</u>. There was something else going on. That is not to say someone might have been on Venus before all this happened and used some underarm spray and got yelled at. I'm going to give you a super-fast overview of what happened up there.

- **Very Recent**-First it was not extremely long ago [about 11 thousand years ago. Scientists have all wondered why all the artifacts they find are very young as the massive volcanic action that send molten lava all over the place happened VERY recently.

- **Equatorial Craters**-Second, most of the major craters are along the equator so most likely something hit the

68

surface that once revolved around it and was nearby. While this usually means a moon was around the planet there is no moon today.

- **Carolina Bays**-Third, 10 to 12 thousand years ago well over 500 thousands meteoric pieces peppered the eastern coast of the United States and the crater evidence is still extant. The craters are in a line as if it was a time when the Eastern Coast was the Equator and as the meteors fell, the Earth rotated.

- **Historical Reference**-Fourth, dozens of ancient text talk about the "Planet with a wavy tail, or Planet with a fiery tail, or angry planet, or vain planet" that destroyed much of the land as if many saw the destruction of Venus during ancient times.

- **Biblical Planet Rahab**-Fifth-The Bible directly discusses the destruction of the Planet Rahab as a military garrison was stationed there during an ancient war and the entire planet was destroyed.

- **Strafing Evidence**-Sixth- a number of in-line, exactly circular and same sized bomb-like blast have been found on the surface of Venus.

- **Biblical War Description**-Seventh-The ancient Biblical Texts describe a horrible War just before the End of the Pleistocene where 1/3 of the entire world population was killed.

- **Pleistocene Extinction**-Eighth-Very soon after the Carolina Bay meteors hit with horrible force, we know the earth axis shifted, the world became unsteady, water from massive tidal waves covered the world and almost complete extinction of life accompanied what we call the Pleistocene Extinction 10 thousand years ago.

I know you are thinking these are only coincidental, and I don't want to provide the massive amount of data that confirms these things here, but I do want to give you some brief data next. First let me give you a few of the Biblical texts describing the Massive war and the destruction of the Planet Rahab/Venus. Over and over again it says the same thing. The Planet Rahab [Vain place] was completely destroyed. Huge pieces of debris were made as it shattered and the pieces became "stones of fire" or meteors.

Psalm 89:10 - "Thou [God] hast <u>broken Rahab in pieces</u>, as one that is slain;" [The pieces sound like meteoritic pieces. Especially as we read further.]

Isaiah 51:9-"O arm of the LORD; awake, as in the ancient days, in the generations of old. Art thou not it that hath <u>split Rahab</u>, and wounded the dragon?" [The Dragon most likely was the leader of some military group. Note the <u>idea that the planet was split</u> as is seen in the topographical map that follows.]

*Job 26:12- "The boastful Angel and his followers rebelled. Yahweh destroyed their dwelling places. He divideth the sea with his power, and by his discretion <u>he smashed Rahab</u>. It was reduced to <u>**stones of fire**</u>."* [By this verse we could well believe that many people had made Venus their home before the disaster.]

Enoch 85 and Revelation 9-"I beheld a single star fell from heaven-then I beheld many stars which descended and projected themselves from heaven to where the first star was." [This could very well be the vision of many meteorites hitting the Earth.]

"Jasher" provides us with estimates of the destruction during the disaster. It indicated that 1/3 of the inhabitants of the entire Earth were destroyed. Here are the specific Biblical verses.

Jasher 2:5-6- *"-and the sons of men forsook the Lord all the days of Enosh [Adam's grandson] and his children; and the anger of the Lord was kindled on account of their works and abominations which they did in the Earth. And the Lord caused the waters of the river Gihon to overwhelm them, and he destroyed and consumed them,* **_and he destroyed the third part of the Earth,_** *and notwithstanding this, the sons of men did not turn from their evil ways--"* [This was a horrible war and Venus got involved. I know you are wondering how they got up there, but right now you look at the evidence.]

Isaiah 14:12- *How art thou fallen from heaven, O "Heylel" [**morning star**] , son of the morning! how art thou cut down to the ground, which didst weaken the nations!* [Essentially, this is talking about parts of Venus falling to the ground and weakening the nations. The idea of weakening nations sounds like something hurting the human inhabitants of the earth such as one would expect from the huge meteorite storm aftermath of an exploded moon of the nearby planet Venus. By the way, this Heylel term is used no place in the Bible except for this verse.]

Craters That Aren't Meteor Craters

The row of craters below left are not indicative of a meteor shower that would cause a random layout of variably sized blasts as the meteor exploded high in the atmosphere. These strikes are directed in a line on Venus. Here are seven blast areas in line. Someone was apparently trying to hit something during a strafing run of some kind. One thing that should be noted is that each of the blast areas is exactly the same size so the blasts could not have been random pieces of meteor unless each piece came from the same source, all happened at the same times and all pieces were the exactly the same size and density. This is indicative of bomb blasts. Another is shown to the right.

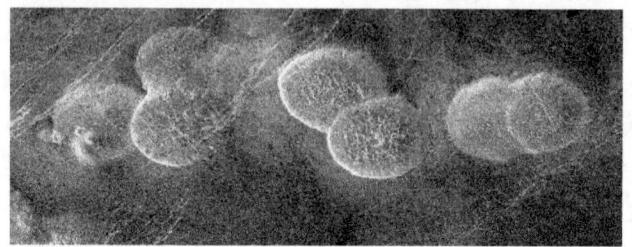 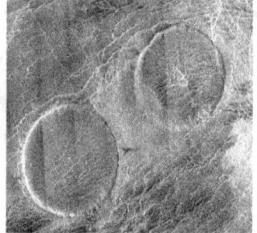

Images of Venus

Next, are a few of the many images showing how it used to be. The remains of massive waterways are still extant, but all the water was quickly removed when the Planet died. Also below is an image of the middle of Venus showing the massive split as the planet almost was cut in two.

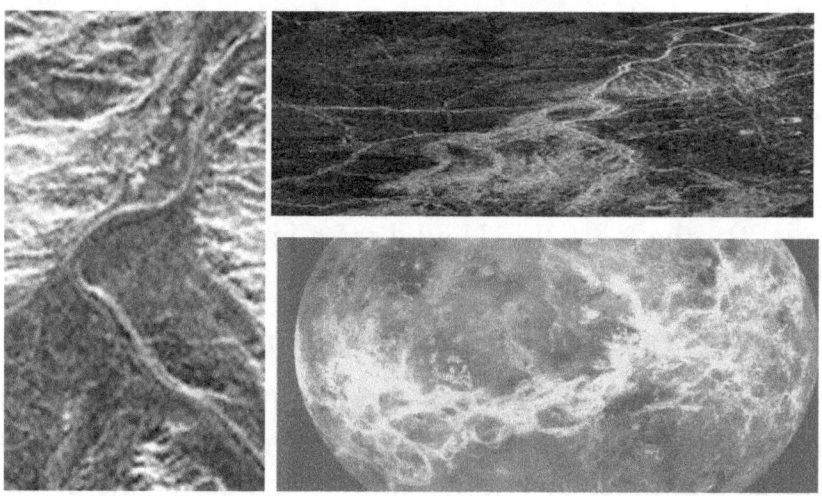

Argon-A main curiosity found by the Magellan probe was that the atmosphere of Venus contains high levels of the isotopes of argon, neon and noble gases. These high concentrations of noble gases could only mean that the current atmosphere of Venus is extremely young [on the order of thousands of years---say 10 thousand] because noble gases don't combine with other materials and escape easily into space; even with a thick atmosphere.

Plasma-To add craziness, in mid-1997, the Soho satellite detected a plasma structure issuing from Venus and it is long enough and in the right direction to almost reach the surface of Earth. The report described the structure as "stringy." Now for the interesting part; such a structure could only remain intact if a current were continuously flowing from Venus to the surrounding space via the plasma tail. "The initiator could very well have been uneven electrical charges between Venus and Earth. This could mean a massive electrically charged plasma burst reached Venus as our planets came too close together. I know that sounds science Fictiony, but Plasmas are used every day in our Fluorescent Lights, so don't discount them and Planets spinning all over the place WILL build up a charge and the chances of 2 planets having the same charge is almost impossible.

One Third of all Men Died in War

Even before the flood, there was massive destruction everywhere. Let's take another peek at Jasher to fill us in on some of it.

Jasher 2:4-7- and the sons of men forsook the Lord all the days of Enosh and his children; and the anger of the Lord was kindled on account of their works and abominations which they did in the earth. And [later] the Lord caused the waters of the river Gihon to overwhelm them, and he destroyed and consumed them, and he destroyed the third part of the earth, And in those days there was neither sowing nor reaping in the earth; and there was no food for *the sons of men and <u>the famine was very great</u> in those days.*

This was not the worldwide flood, but is a recognizable flood and devastation time identified around the world. We know, by several other texts and physical evidence, a huge flood overtook the world about 120 years before the end. A second even more devastating one occurred about 10 thousand years ago. This was the first flood.

While it initially looks like the rivers overflowed in the time of Enosh's children [the time of Adam's great, great, grandchildren], other verses enhance the description and indicate that it was really during the time of Jarad [Enosh's grandson], that the horrible times were near their peak. We are also given specific numbers as <u>1/3 of the world's population died</u> in this flood before the worldwide flood following. Just imagine! ---This was before almost everyone died in Noah's flood.

Enoch 106:13-16. *And I, Enoch, answered and said unto him: 'The Lord will do a new thing on the earth, and this I have already seen in a vision, <u>and make known to thee that in the generation of my father Jared some of the watchers of heaven transgressed the word of the Lord</u>. 14. And behold they commit sin and transgress the law, and have united themselves with women and commit sin with them, and have married some of them, and have begot children by them. 17. And they shall <u>produce on the earth giants</u> not according to the spirit, but according to the flesh, and there shall be a great punishment on the earth, and the <u>earth shall be cleansed from all impurity</u>. 15. Yea, <u>there shall come a great destruction over the whole earth,</u> **then** <u>there shall be a deluge and a great destruction for one year</u>. 16. And this son who has been born unto you shall be left on the earth,*

74

and his three children shall be saved with him: when all mankind that remain on the earth shall die. And all the idols of the heathen shall be abandoned, And the __temples burned with fire,__ And they shall remove them __from the whole earth__,

Notice in this section of the book of Enoch that there is a great destruction just before the great deluge. This is referencing the Meteor Storm that will lay hundreds of thousands of craters on the East Coast of the United States. A few hundred of the thousands will be described after the war information.

What Happened To Venus

Let me tell you what I believe happened. I wasn't there and I don't think they had too many people spraying Freon, but what they did have is a close distance to the Earth and one time it got too close, just like Mars did many years earlier. This time, a plasma string [still seen today] connected the 2 planets causing a huge electrical surge as planets spinning around the sun build up highly different potentials between each other. This is not a problem unless they get too close.

The electrical "lightning bolt" must have hit the moon of Venus and it shattered. The close proximity of the destruct made many craters along the equator of Venus and hundreds of thousands of pieces headed in the direction of the plasma from Earth. Fires ignited the Earth, but Venus was in much worse shape as its rotation slowed too much and the thick atmosphere got so thick the surface was ignited. Lake, rivers, cities and everything else was destroyed in an instant as the temperature rose to 700 degrees and melted people stationed there. This would end the war and end Venus as a livable "Vain Planet".

I guess you could call this global warming, and it could happen here, but it is highly unlikely.

Proof of the War

Just before this destruction is a time we find something called the Young Dryas. While Dyra is a flower, the time was no sweet smelling loveliness. Instead it was a wild time and it took a long time; on the order of a thousand years. Imagine a 1000 year war! It was the end of the last Ice Age. Temperatures increased steadily for the next thousand years, but this is where it starts getting interesting. Levels of copper, tin and lead show marked increases and there is this abrupt ending that looks really weird which has a huge drop in temperature as measured in Greenland.

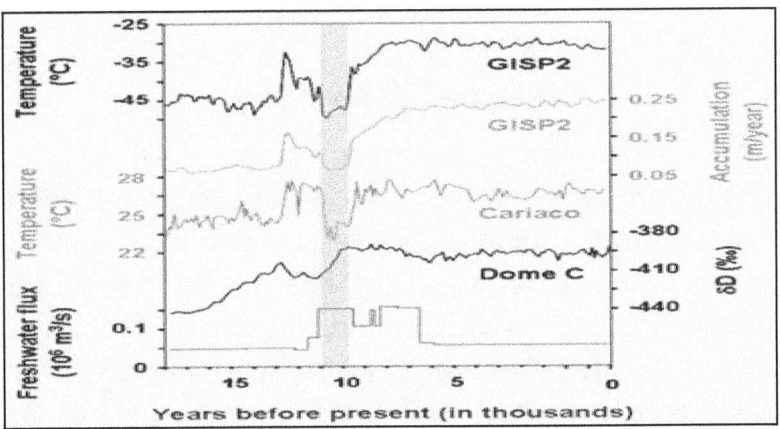

Death was everywhere and then we find something very scary. **Uranium concentrations in coral jump by almost 300%.** Also we find marked **increases in nanodiamonds**, magnetic spherules [tiny balls], and **carbon spherules** at

the end of the War with a **major increase in charcoal**
around the middle showing fire unbearable heat and
nanodiamonds indicating nuclear explosions confirmed by
the uranium concentrations found. ----- This occurred
during a fairly brief time in our history as shown next as
depth can be interpreted as time.

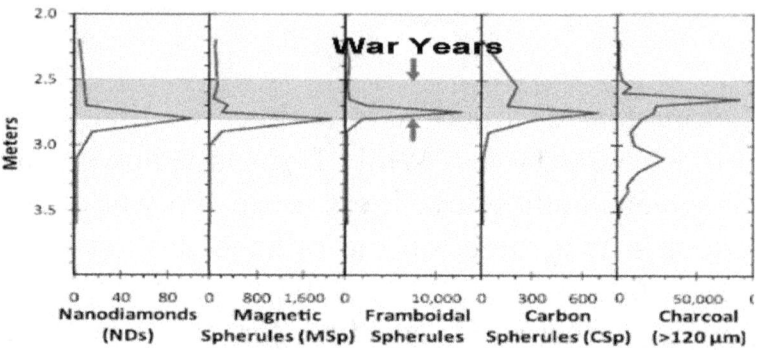

Miles of fused desert sands in Libya and Egypt may also
show a massive high temperature "explosion" without a
meteor. All the evidence seems to support the existence of
an ancient nuclear war having taken place at this time and
felt around the world.

Xenon-129 Evidence

One way to test for nuclear explosions is by looking for
radioactive remains such as Xenon-129. Xenon-129 stuff is
a "second order nuclear fission by-product" and guess
where too much is found. Mars has nuclear by-product in
abundance. Forget I said anything about this, bringing up
Venus was bizarre enough. Let me get back to Earth and
see what has happened in North America and along the
coastline of the United States in particular as this coastline
had be along the equator during the war.

Carolina Bays

In North America, something similar happened and quickly halted the world war that Jasher indicated cost the lives of 1/3 of the population. The huge particles of the Venusian moon blast traveled quickly to the Earth and many hundreds of thousands of massive particles hit along the coastline of North America and around to Australia. Today 500 thousand plus craters up to 14 miles across are still extant. The impact was enough to shift the Earth axis and only those in sealed ships or flying machines survived the massive 500 mile and hour winds, mile high tidal waves, a month and a half of rain, earthquakes and volcanic action around the globe. After the Earth settled, huge piles of animals were left to rot and the cities were all destroyed. The image to the left show a small sampling of a few hundred or so of these craters that are everywhere! To the right shows the blast area all in a straight line of this coast and even in Australia all happening at the end of the Pleistocene.

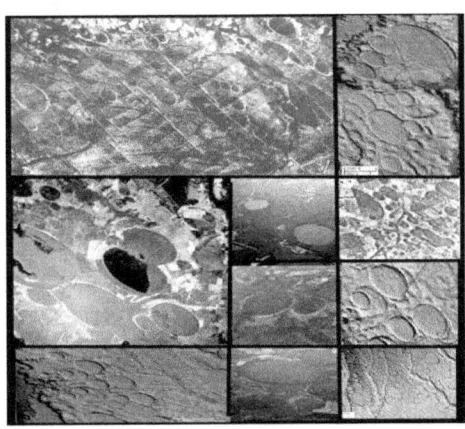 

The meteors could have initiated the climate change and eventual axis shift that caused the worldwide flood, but whatever it was, soon the Earth was in serious trouble.

120 Years Before the Flood

We may even know how long it took for the destruction to be completed and the Earth shift that caused the end of the world by water.

Genesis 6:4-And the LORD said, "My Spirit shall not always strive with man, for he also is flesh; yet his days shall be <u>a hundred and twenty years.</u>"

Jubilees7:19-20- And He said 'My spirit shall not always abide on man; for they also are flesh and their days shall be <u>one hundred and twenty years</u>'. And they began to slay each other till they all fell by the sword. <u>And the Lord destroyed everything from off the face of the earth;</u> because of the wickedness of their deeds.

Jasher 5:9-11 And Noah and Methuselah spoke all the words of the Lord to the sons of men, day after day, constantly speaking to them. But the sons of men would not hearken to them, nor incline their ears to their words, and they were stiff-necked. And the Lord granted them a period of <u>one hundred and twenty years,</u> saying, If they will return, then will God repent of the evil, so as not to <u>destroy the earth.</u>

It looks like the destruction of Rahab was a warning. After 120 years, the world ended. Scientists call it the end of the Pleistocene Age. Many tests of the water height of the Atlantic shows that during this time, the water level increased by 300 to 500 feet and NEVER returned to that level. The end of the Pleistocene was upon us.

After Extinction

> *With all of this, the Earth did not experience thermal runaway.*

We don't know all the details in the horrible wars and massive loss of life, but the worst was still to come as the Earth flipped on its axis by 30 degree placing Siberia in the Arctic Circle and all the Mammoths were gonners. Torrents of rain and violent earthquakes and the equatorial shift quickly melted the previous polar regions filling the world with water. A few were saved by finding secure vessels to stay in as the earth shift cause another monster. Unbelievably high tidal waves crashed over the land, drowning all who remained that were not protected, according to the Biblical testimony, as well as 80 thousand other ancient texts describing the horror. Tiny segments of people survived around the world and very quickly, civilizations were advancing at tremendous rates. We can believe factories spewed out noxious gases as the world was reviving. The time was 10 thousand years ago and the earth had not succumbed to global warming.

Another Massive War

Unfortunately, the tensions of the world were not settled long as power struggles soon erupted and within another 4 thousand year, the entire world was again at war. We know the time and destruction of the war by writings, physical evidence, scientific studies, and cross comparative analysis. The war ended around 3100BC. The Egyptians called it Zep-Tepi [new beginning] the Maya started a new calendar; those in India described it as the new Age of Kali. While some called it the Bharata War, the ancient Jewish texts called it the Tower of Babel War. The book of *Jasher* told us **1/3 of the entire population of the world was lost**, another 1/3 of the population became like apes and the rest scattered to far reaches of the earth. I'm telling you all this to show you that massive increases and decreases of world populations and hydrocarbon production has cyclically been going on for thousands of years.

Fear Mounts

If people had anything to do with the global temperature variations, we would have noticed the change during this last Great War and all the rest that nobody told you about. With that, let's look at what you ARE being told with a little more detail!!!!!!! For one thing, they do not tell you about the details of these ancient conflicts that certainly pushed substantial greenhouse gas into the delicate atmosphere. Another thing that is not indicated is that there were so many people living in the world during these times, people COULD affect the atmosphere. Finally, what they did see, they have misinterpreted and caused those who go against them great hardship. I'm not saying all those believing in the near term catastrophe of global warming by mankind are all deceitful, power hungry, people trying to introduce an artificial need for some special interest industry that they have massive interest in. What I am saying is *always follow the money*. If coal is being targeted; what industry and what people might gain power or money.

I'm all for altruism and even directed efforts to truly save our planet, but there do not seem to be many of the leaders in this calamity who are sincere. It truly is a shame!

New industries for Nuclear Power, Wind power, and Solar Power, were competing again the Oil barons and Coal to support comfortable life styles around the world.

Governments were pressured into providing "special incentives for these industries in a race to halt the deadly "Global Warming" menace.

Nuclear power seemed to be well placed to be the leader, after the huge cost of putting in plants, the running cost was even less than that of coal, but problems soon arose, as accidents around the world put another fear into people and the inability of anyone to find a place for the "dirty water" effluent allowed more inefficient energy producing components to surge ahead. Please notice Geo-thermal and Gas are about 2 times as expensive as coal while wind is 3 times and solar is 5 times. I know you were told this when you got your solar water heater, but subsidies are trying to artificially lower costs to get people used to them.

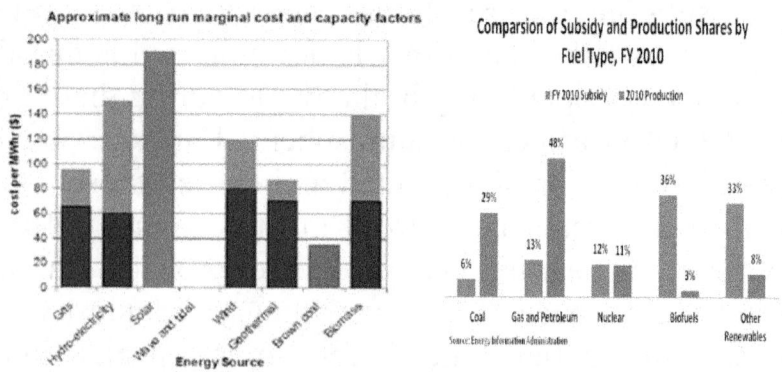

As shown to the left huge susidies are being provided for everything besides Coal Gas, and nucleaer energy with almost no useable production as the inefficiency cannot be eaisly overcome.

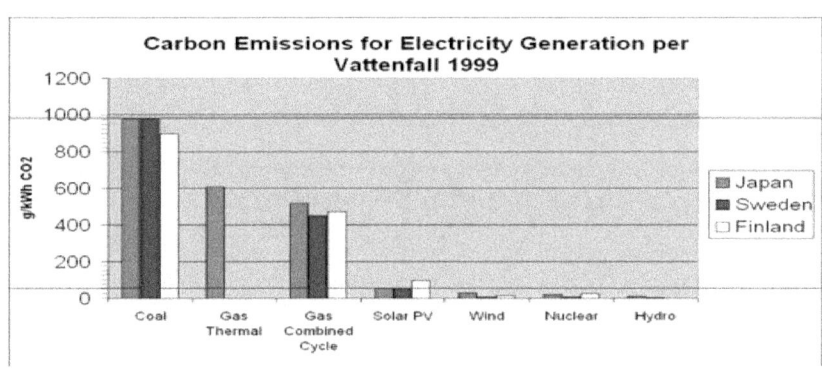

The preceding chart shows almost all electricity is still made from coal or a combination of coal and gas. If all the hydro-electric, Nuclear, Solar, wind and all other electricity producing methods were destroyed, we would see almost no effect [except for fall-out from the nuclear plants]. Therefore; we need to be somewhat sure that eliminating coal will save us. After all; until 1990, we produced more coal than Asia, Europe, Soviet Union, Australia, South America, or Africa. Today, we are still the 2nd largest producer, but China has taken a huge step in energy independence as it increased coal production by 500% since 1990.

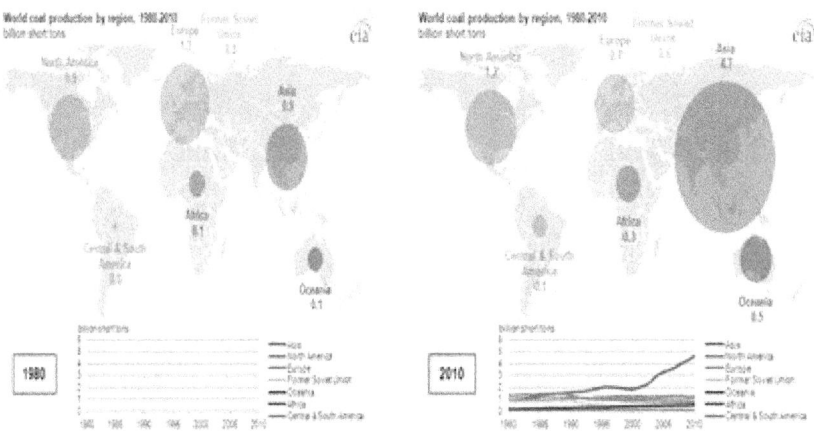

We will discuss the vilification of Coal and CO_2 as the greenhouse knock our chemical, but what about wind?

Wind Power

As shown below, Wind Power capacity in the United States has increased by leaps and bounds or 1900% from 2000 until 2011. One would think that wind power was the savior of the planet. As shown previously, almost all this energy is so greatly subsidized, it is crazy to continue, but someone is making big money----saving the world. Let's make sure he is really saving it.

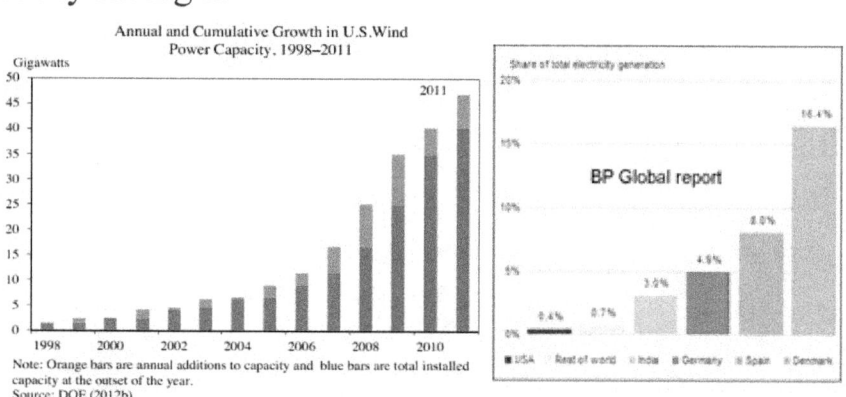

Annual and Cumulative Growth in U.S.Wind Power Capacity, 1998–2011

Note: Orange bars are annual additions to capacity and blue bars are total installed capacity at the outset of the year.
Source: DOE (2012b).

The chart to the right shows that even with this enormous capacity growth, there is very little energy produced as little old Denmark produces 4100 percent more energy this way than the massively subsidized United States.

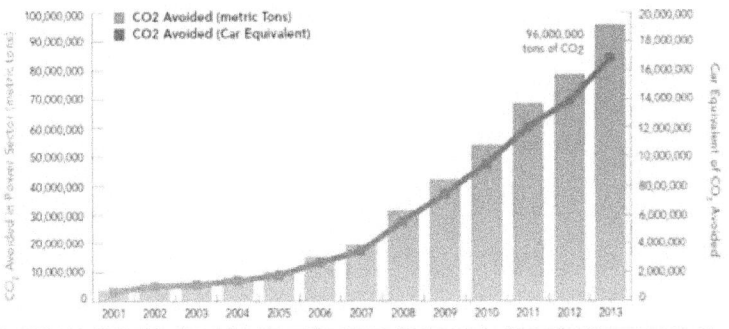

Avoided CO2 Emissions from Wind Energy

- In 2013, wind generation avoided an estimated 95.6 million metric tons of carbon dioxide (CO2)— the equivalent of reducing power-sector CO2 emissions by 4.4%, or taking over 16.9 million cars off the road.

- The 12,000 MW of wind power capacity under construction at the end of 2013 would reduce another 20 million metric tons of carbon dioxide (CO2) when it is operational — the equivalent of reducing power sector CO2 emissions by another 1%.

I thought it would be neat to show how these guys perk interest. According to this AWEA [Wind Industry] report. The world has saved 96 million tons of CO_2 getting into our air since 2001. Let's see if Solar industries are also doing their part and cashing in.

Solar Power

While wind is a travesty, solar power generation has been an utter failure as the following graphs shown. Soon it was the biggest cash cow for all sorts of unscrupulous Congressmen, Business Executives, and Scientists in the NOAA and similar organizations. The first shows solar energy subsidies cost 150 times as much as the laughable wind power pursuits and **121000 percent** more per megawatt than coal. The second one is even worse. It shows that our government is spending 200 percent of what the wind power guys get to 0.007 times as much energy as coming from the already inefficient wind power.

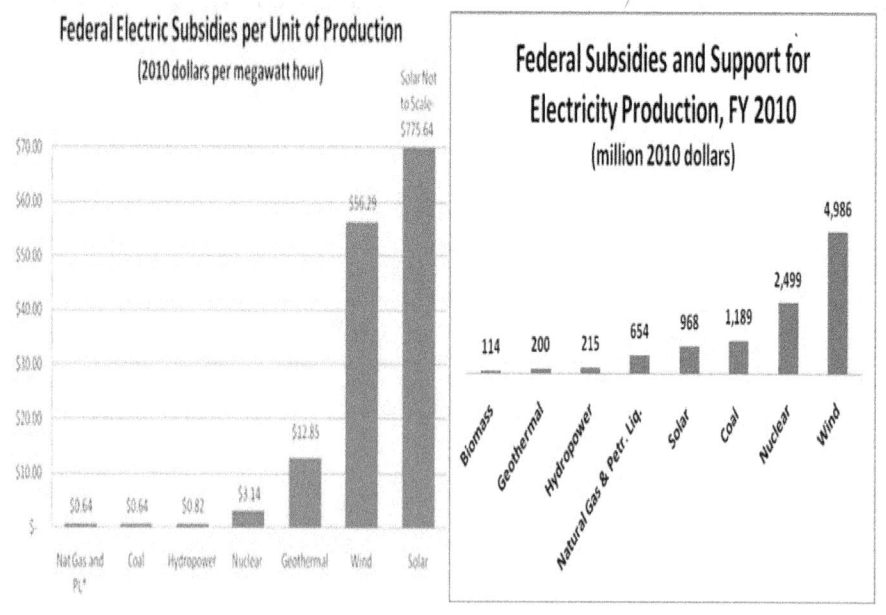

Unfortunately for you and me the fear mongers have told everyone that CO_2 was killing the Earth. Pressure from those being misdirected allowed awfull legislation to allow many to gain huge wealth art the expense of taxpayers. All they had to do was to fabricate a horrible calamity and we are paying the awful price. To make this worse, many of the companies gaining these subsidies are foreign with the highest concentration from China.

NOAA Disception

Luckily, we have an agency looking out for us called NOAA [National Climate Data Center and National Oceanographic and Atmospheric Administration]. Sorry that was a joke. There is no question that these people are NOT. What we thought was 100 of the Top Global Thinkers [according to the *Bulletin of the American Meteorological Society*], but what we now know is they would do ANYTHING to advance their AGENDA.

First Indication- ARGOS

Scientists designed 3000 ARGO buoys that just floated around and took temperature measurements since 2003. These buoys show absolutely NO thermal increase [but the published data from NOAA somehow showed massive changes that the buoys "SOMEHOW" missed. The NOAA team decided that the information from them should not be used. One reason noted is that they weren't floating near the Arctic. Some might wonder why NOAA would have helped place these things and later decide they were stupid.

Some begin to suspect that there was a large network of politicians, corporations, and scientists that were conspiring to promote the fear of "global warming" . . . despite evidence clearly stating no such "global warming" exists. With only $22 billion being pushed into the global warming epidemic, you might wonder why some would try such a scare tactic.

NOAA Published Mistake

The National Climate Data Center and National Oceanographic and Atmospheric Administration [NOAA] put out a chart showing there was no significant temperature rise in the United States from 1940 until 2010 and that the spring was the coldest in the 115 year record, but at the same time they told everyone to ignore this data and focus on eliminating the Coal Industry.

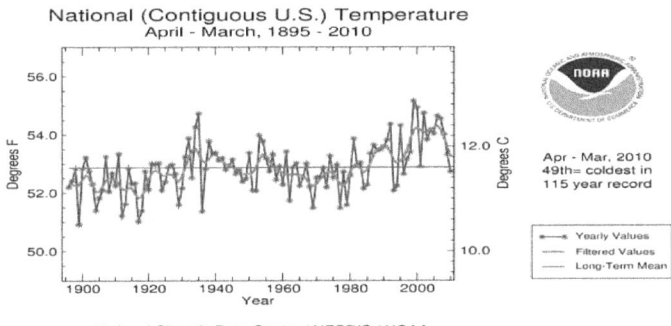

After realizing their mistake they put out a completely different chart so that people would fund their pet projects associated with Earth annialation. By focusing on the slight rise since about 1960, the danger begins to look real. We are going to look into this chart that has been the foundation for many, many others and is still being used even after the fraud was fully exposed last year. You will be seeing this same massive thermanl ramp again and again, but I will show you the true slope for the original data from the Bouys and sattelites.

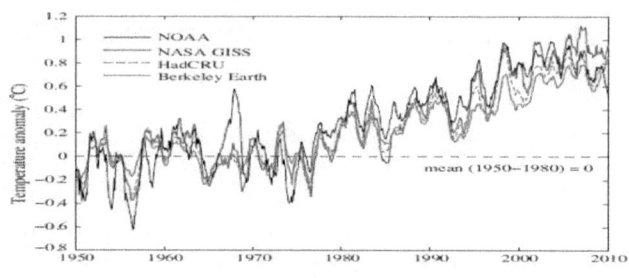

17 Year Cooling

NOAA got fancy with this one. According to NASA's own data, the world has warmed 0.36 degrees Fahrenheit over the last 35 years, starting at the fairly cold year-1979. Even this would show a massive increase of 0.1 degree per year over that short time. The NASA Remote Sensing Systems data also shows that since 1998, the average temperatures around the world have been steadily decreasing as shown below.

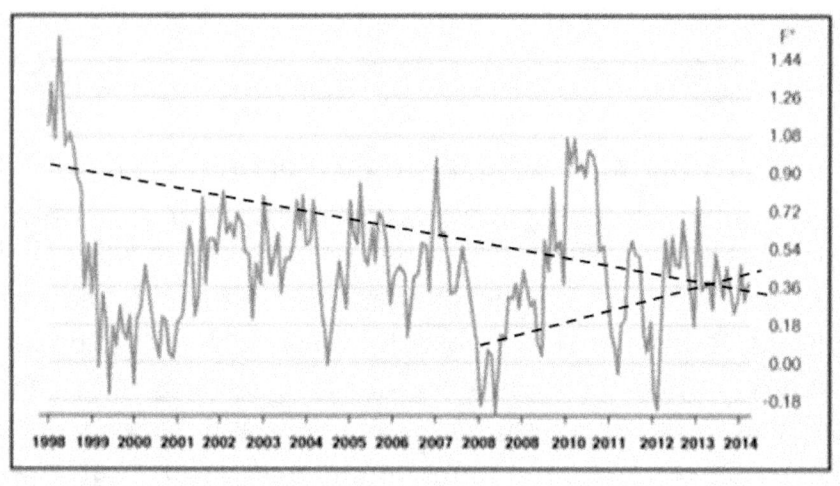

Source: NASA, NOAA and Remote Sensing Systems

According to this graph, the world is 1.08 degrees cooler than it was in 1998. NOAA has not used this data when "Informing the World" about the condition of our Earth. To make it look like horror, just truncate until a rise is shown [2008 until today shows a 0.32 degree rise]. This is not the new data I mentioned to correct the previous longer term graph. That is next.

Just Plain Lying

NOAAs current US graph is shown below left [same as before]. *Now we know it is all a lie*. Note that there is a discontinuity at 1998, which doesn't look right. <u>Globally, temperatures plummeted in 1999-2000</u>, but they didn't in the US graph. Note that measured data below right shows that by 2008, temperatures were back down to the 1989 level. But in the NCDC data, 2008 is half a degree warmer than 1989 making the temperature LOOK like a disaster when there is almost no change at all. Please note that the faked chart to the left has been used to justify an enormous number of "charts showing the destruction of our world.

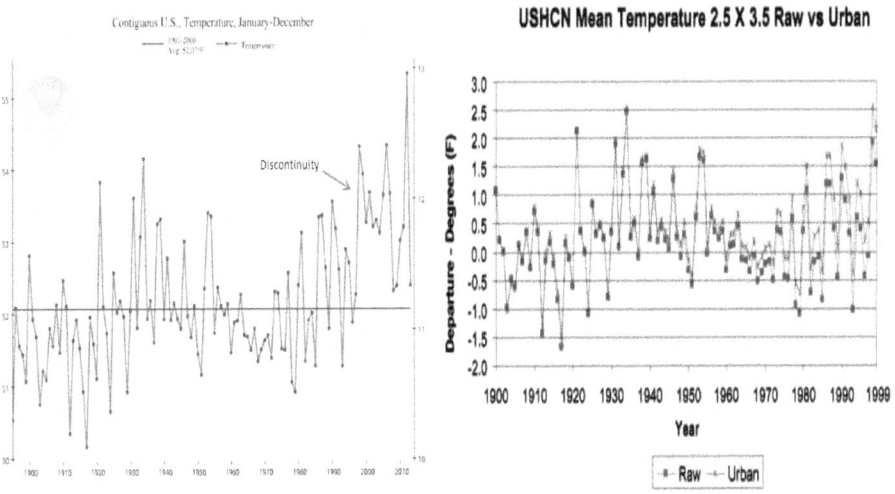

By putting the 2 togerther we can easily see the treachery. The top graph is from RAW data and the bottom one is the "doctored" chart making everyone want to give money to Green Technologies to protect them from this FAKE temperature rise.

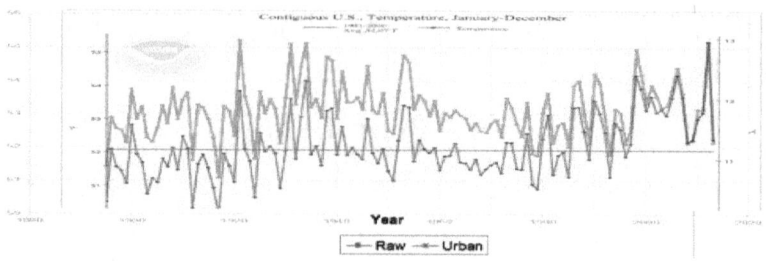

Bottom line is that NOAAs US temperature record is completely broken, and meaningless.

Adjustments that used to go flat after 1990 now go up exponentially. Adjustments which are documented as positive are implemented as negative to amplify fear for monetary gain.

1940 Spike Removed

The best way to describe the subterfuge is to talk about the 1940s heat wave. This was a huge thermal "spike" occurring as the Earth recovered from the mini-Ice Age. If you were somewhat devious, you would like that spike to go away as it shows the temperature today is not significantly different than 1940 levels. This can be done 2 ways.

1. Show the thermal rise since 1965. This is the favorite one.

2. Change the 1940 peaks to smooth out "anomalies that don't go along with the "THREAT". In this case it seems unthinkable, but this is another way to make people buy Solar Cells for their homes. [Especially when the US Government pays for most of the installation to eliminate the use of Coal.]

We now know "cooling the past" adjustments have been carried out in the Arctic region to an unbelievable level of sabotage. Nearly every current station from Greenland, in

94

the west, to the heart of Siberia has been altered in this way. The effect has been to remove a large part of the 1940's spike. I think the chart below left shows you how it helps breed fear. The actual data has high temperature bumps centered in 1940 and 1960, but the common practice is to eliminate the annoying bumps to make the temperature seem to be ----OUT OF CONTROL. Sometimes a little of the spike data is kept in but the density is reduced to show an average increasing slope is more terrifying as shown below right.

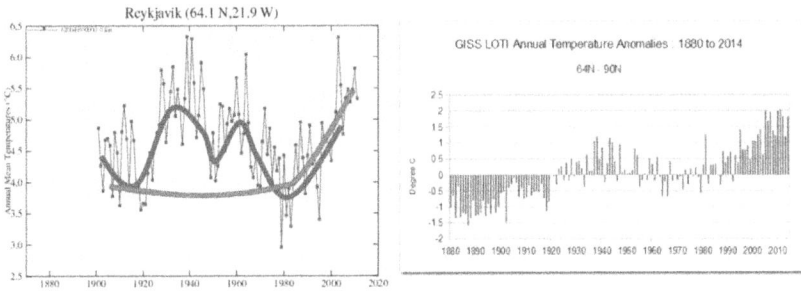

After the practice was revealed from some cleaver undercoverwork and retreval of internal EMAIL traffic, we have started to see a few of the practicioners coming clean about their part in the underhanded fear mongering.

Scandalous Actions

Called *Mike's Nature Trick*, perpetrators of the change in temperature charts that greatly fed the fire on global warming fear are now known and the act has been proven. It seems none went to prison or were even fined for what is perhaps the worst type of criminal extortion and conspiracy. While generally these guys are protected by those giving unbelievable wealth, the sad fact is there is no specific law for holding the world hostage to fear and with that fear being able to extort BILLIONS of dollars from the innocent.

Coming Clean

Dr. Phil Jones — a leading "global warming" advocate at the United Nations — admitted that he used *"Mike's Nature trick" in a 1999 graph to "hide the decline" in temperature.* He was never charged with any crime--- nor was the originator of the deception.

Dr. Stephen Goddard had this to say: *"The National Oceanic and Atmospheric Administration (NOAA) has been "adjusting" its record by replacing real temperatures with data "fabricated" by computer models."*

Dr. Robert Stavins, who helped write the 2014 United Nations Climate Report, revealed that *"politicians demanded he change and edit parts of the report to fit their needs!"* You would think his changes would land him in prison, but he is still one of the "respected" scientists.

Dr. Kevin Trenberth — One of the NOAA elite finally admitted, *"The fact is that <u>we can't account for the lack of warming</u> at the moment and it is a travesty we can't."* If there is no warming all of the billions of dollars going to the select would dry up and the United States could spend money helping the majority of Americans instead.

D. James Lovelock, was awarded the Wollaston Medal for his climate work in something he called Gaia Earth. He was one of the first to proclaim that *humanity would soon end due to global warming from CO2.* After the last 15 years with no global warming, he has admitted to being "alarmist" about climate change and says-*"Other environmental commentators, such as Al Gore, were too...The problem is we don't know what the climate is doing. We thought we knew 20 years ago. That led to some alarmist books – mine included – because it looked clear-cut, but it hasn't happened."*

Al Gore Gets Rich

Al Gore has done a tremendous disservice to humanity and gained millions of dollar doing it. I'm talking about the 2007 Nobel Peace Prize winner who gained fame in his push to place fear in the hearts of all concerning the Faked" global warming scare we help push.

His "Ice" Lie

This guy stated the following, *"The North Polar ice cap is falling off a cliff. It could be completely gone in summer in as little as seven years. Seven years from now."*

The images below show the Ice cap in 1990, 2007, 2013 and 2015. From the minor reduction in 2007, the Ice caps have been almost steadily increasing to over 63% larger.

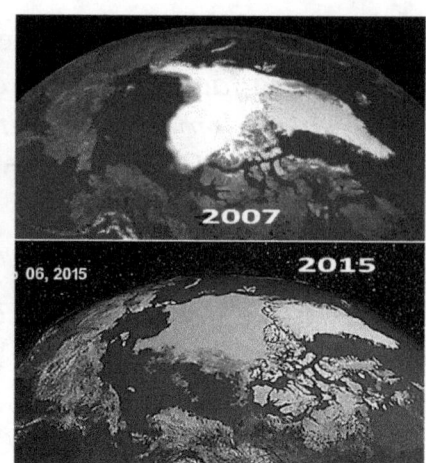

"Man Caused Death" Lie

Al Gore claimed *CO$_2$ emissions from Human factories were destroying our world as massive amounts of factory effluent were making the Earth's temperature go out of control and* <u>*97% of scientists agree it's real*</u>*, it's man-made, and it's dangerous.* Certainly, he knew the satellite data and all the rest, but he was making a fortune. He also knew many scientists were begging for people to listen to them as Mr. Gore misrepresented everything.

I know it seems like everyone you meet indicates that man-made green gases are killing the earth and there are charts showing that most scientist agree and thanks to a fabulous campaign of deceit, the "normal population" now believes we are destroying our planet, but Please just read the book and then make up your mind rather than blindly listening to some reporter on CNN or the president of some Solar Cell manufacturer tell you how much less Carbon is being introduced by giving up underarm spray or taking a bicycle to work or whatever. Have you ever heard the term "follow the money"?

Give Me Money

Al Gore, was worth $2 million when he left the Vice Presidency in 2001 and now, after investing almost entirely in a small number of "GREEN-TECHNOLOGIES" raping governments out of fear, Gore is worth $100 million and it is growing as fast as green-technologies can go. The massive artificial funding of these horrible industries goes away if global warming is not pumped up, so Gore seems to do whatever is needed to make his "investments" thrive when there is no way for them to thrive. Lying, misdirecting, lying some more, it's a business of fear.

Scandal After Scandal

While there are many "me-toos" associated with Al Gore, let me talk about one we know fairly well. His name is Barrack Hussein Obama and he has done more to cripple our natural resource independence of any President in my opinion. While never getting a Nobel Peace Prize for Global Warming, he certainly has been doing his best to excite our population. He tweeted last year the following in support of Al Gore's statistic: *"97% of scientists agree: climate change is real, <u>man-made and dangerous</u>."* Thank you Mr. President. Let's just destroy the coal industry right now just like you promised the Europeans you would do. Luckily, the Wall Street Journal reported the following, *"The assertion that 97% of scientists believe that climate change **<u>is a man-made, urgent problem is</u> <u>a fiction</u>**."* When further review was done, it was discovered that many did believe the earth was getting warmer, but a mere 1% of scientists believe <u>human activity is causing most of the climate</u> change. Unfortunately, that was in 2006. Today that has changed as most seem to have been brainwashed into believing this horror.

Not to fear, a petition was signed by more than 31,000 "regular" scientists that states *There is no convincing scientific evidence that human release of . . . carbon dioxide, methane, or other greenhouse gases is causing or will, in the foreseeable future, cause catastrophic heating of the Earth's atmosphere and disruption of the Earth's climate."*

The popularity of fear has worked its way into every area of our society and President Obama increased the fear by saying, *"We want our children to live in an America that isn't threatened by the destructive power of a warming planet."* It is so criminal what has happened. I know my little book isn't going to help much, but hopefully some will begin to see more of the lies and stop some of this madness as we try our best to bankrupt our country on crazy scheme after crazy scheme. Money is obtained by many ways but the most insidious is by something called loan guarantees.

Relying on Global Warming

The current Department of Energy failed Loan Guarentee Program is split into 3 types of loans as shown in the graphic below, but all require the public fear of global warming to keep the failed industries from having any chance at paying some of the money back and not forcing a ressession era.

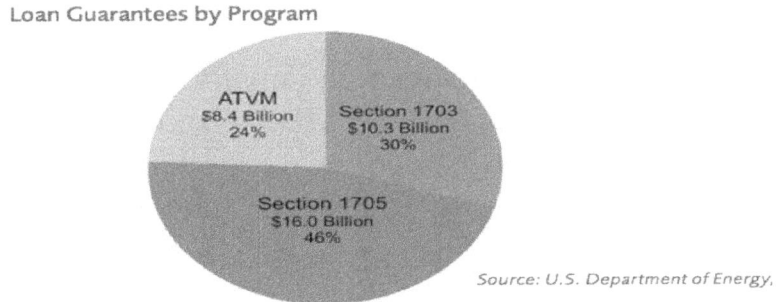

Loan Guarantees by Program

ATVM
$8.4 Billion
24%

Section 1703
$10.3 Billion
30%

Section 1705
$16.0 Billion
46%

Source: U.S. Department of Energy,

Advanced Technology Vehicles Manufacturing [ATVM] dollars were secured fearing gasoline vehicles would turn our planet into a deathtrap. Recovery and Reinvestament

Act Section 1705 was to reduce effluent from bad industries that would make our planet die and build substitute capabilities. Section 1703 seems to just be a way to prop up Solar Cell manufacturing no matter what the cost. Forget scandal, this mess has turned into calamity. The visible dollars are shown as 34.7 Billion, but the total cost is an extremely difficult to estimate and consumers are repaid to put in solar, geo-thermal, wind and other components to prop up these fear built industries. Section 1703 was almost totally for solar power generation with one company NRG Energy receiving almost $4 Billion alone, with Abengoa receiving about $3 billion.

These don't include the Department of Treasury grants under something called 1603 Grants which went to the same companies and other grants under "Recovery and Reinvestament Act funds [Section 1705]. Quoted in the *New York Times* recently, NRG's chief executive, David W. Crane, explained how his company and its partners have secured $5.2 billion in federal loan guarantees, plus hundreds of millions in other subsidies for four large solar projects.

"Solyndra" Scandal *[Solar Cells]*

Obama's Solyndra green energy initiative cost taxpayers $500 million. He, in essence, *used taxpayer money to finance his re-election campaign by funneling it through Solyndra.* He received $1.25 million in campaign contributions from Goldman Sachs and George Kaiser who took over Solyndra when it failed, so our President insured $500M would be returned for that campaign donation. Solyndra not only gained the $500 million 1705 fund guarantees but also multiple millions from the Ex-IM Bank to help sell the company.

"First Solar" Scandal *[More Solar Power]*

Here is a funny one. We gave "First Solar" multiple millions in "loan Guarantees then we loaned one of their subsidiaries $192.9 million to buy First Solar panels from themselves. That wasn't enough so they were given a $16.3 million loan in 2010 and another $646 million in 2011. Then they got another $547.7million loan from Ex-IM bank to subsidize panels outside the US. Somehow they received $4.5 Billion in government loan guarantees, and were able to give almost $200,000 to Democratic campaigns. Where did they get the contribution money?

"Abound Solar" Scandal *[Solar Cells]*

Another scandalous pay-off was to Abound Solar. Obama got a federal loan of $400 million for Abound Solar, in 2010. Just like the others, it went bankrupt after Patricia Stryker got the money she "needed". She contributed *$500,000* to the *Coalition for Progress* and another $85,000 towards Obama's inaugural committee as her bankruptcy payout was huge.

"GE" Scandal *[Wind Turbines]*

GE makes wind turbines and receives tens of Millions in what are called "Green Energy Credits". Because they received all these "credits" and saved our world, they didn't have to pay taxes at all in 2011.

"A123 Systems" Scandal *[Lithium Batteries]*

A123 Systems, a nanophosphate Lithium Battery source supplier, gave their lobbies $1M to get $280M Federal Assistance. Some might wonder if they only needed assistance to pay for the lobbyist???

"Johnson Control" Scandal *[More Batteries]*

Johnson Controls **"won"** a $299 million stimulus grant to build two battery plants, but, instead, it is running only one plant at ½ capacity and pocketing the rest of the money.

"Fisker Automotive" Scandal *[Electric Cars]*

Fisker Automotive was having the usual "can't make a profit and still keep all those giving me money happy blues so the DoE lent them $529 million. Lucky for us, it got so bad in 2011 the DoE halted the money transfer at $193 million.

Harry Reid and "Ormat" Scandal *[Geo-Thermal]*

Retiring in 2015, one scandal that Harry Reid is running from was reported on recently in both the *Washington Free Beacon* and *Courthouse News*. Ormat Technologies owns and manages geothermal plants in California and Hawaii. Reid and various others got them $136 million in economic stimulus funding. The executives essentially stole the money and didn't even present new green technologies at all. As reported in the Free Beacon, *"Reid bragged about securing Ormat another $350 million loan guarantee from the Department of Energy"*. For some reason, Ormat executives have generously supported Reid with donations for his election campaigns and causes.

$22 Million Or Is It More?

We could go on and on and on and on. The fear generated by the NOAA criminally imaginative "scientists" with their creative redistribution of temperatures to make our world seem hotter has gone a long way in reducing money we could have spent on industries with true potential. There is a figure typically associated with Green Energy distributions to scientists, agencies, and Stock Owners, but $22 billion is just what is *spent* on these "global warming" initiatives.

According to Forbes, the total cost to the citizens of the United States is really $1.75 trillion annually. The U.S. Energy Information Administration says- *the green Energy regulations alone could ultimately cause gasoline prices to rise 77% over baseline projections, send 3 million Americans to the welfare line, and reduce average household income by a whopping $4,000 each year.*

Not to be ignored, the Congressional Budget Office [CBO] estimated *the default on DoE nuclear loan program at 50%,* but guarantees the loans anyway. This allows banks to gain profit if the business does survive and the American public to have all the risk and loss.

Stop Crying! We can make up the wasted $1.75 Trillion other ways like not helping the Americans who lost their jobs or small business or the things than made America great.

Modeling Disaster

Besides the outright lying, and taxpayer money to Green industries to make them seem reasonable, If you want to scare someone, use a model showing "Expected disasters". I'm not talking about one or two. My last count was that 73 of these things were being touted as gospel. Don't get me wrong. These are important, the problem is they all failed and caused more alarm and inappropriate action as the forward vision was twisted.

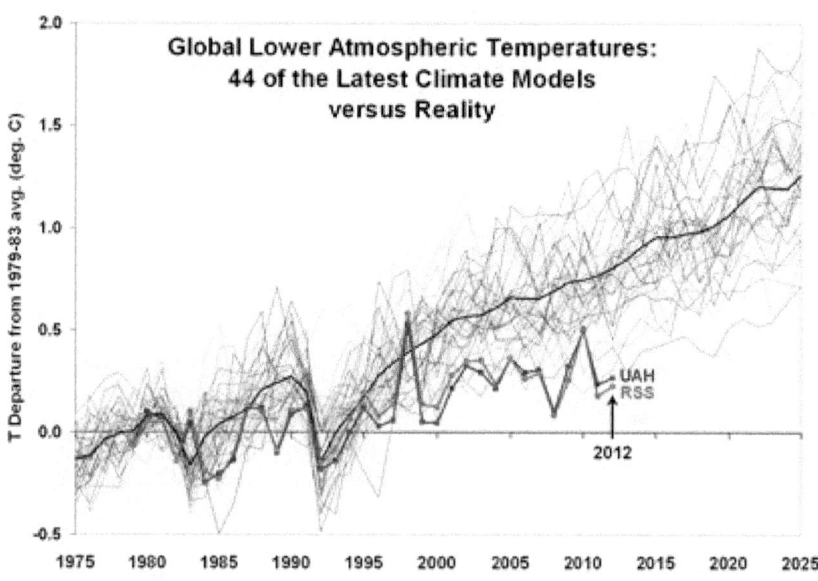

As shown above, 44 of these things are compared to actual Seattleite temperatures through 2012. The ONLY time the 'predicted' Mean World Temperatures were close to actual measured mean values was in 1979 – 1982. The actual

106

temperature rise "estimated" from 2005 to 2012 was <u>250% higher</u>. None of the models is accurate. The trend as shown indicates that the predictions will get worse and will increase in inaccuracy. Clearly, the selected testing elements are not correct. Let me say this once more. The attempts are valiant and models are needed for prediction and research, but we just don't know what causes weather. For someone to blankly state that burning coal will kill us all with nothing to back it up is almost criminal. The actual measurements of mean temperature clearly show that temperature decreases with increasing elevation and these inappropriate and dangerous models do not. The frustration is well noted among those calling themselves Climate Scientists. Even James Hansen tells of his frustration about not understanding weather.

Robert Kaufman, *climate scientist: "...released a modeling study suggesting that <u>the hiatus in warming</u> could be due entirely to El Niño and increased sulfates from China's coal burning."*

Martin Wild, *climate scientist: "During the 1980s and '90s, the rapid decline of air pollution in the United States and Europe dominated the world's aerosol trends. While those emissions have continued to decline in the West, returns, from a brightening standpoint, ... "<u>It's not an obvious overall trend anymore,</u>..."*

Susan Solomon, *climate scientist: ""What's really been exciting to me about this last 10-year period is that it has made people think about decadal variability much more carefully than they probably have before," ...Solomon had shown that between 2000 and 2009, the <u>amount of water vapor in the stratosphere declined by about 10 percent</u>. This decline, caused either by natural variability — perhaps* related to El Niño — or as *a negative]feedback to climate change,* likely countered 25 percent of the warming that would have been caused by rising greenhouse gases..."

Kenneth Trenberth, climate scientist: "*Until 2003, scientists had a reasonable understanding where the sun's trapped heat was going; it was reflected in rising sea levels and temperatures. Since then, however, heat in the upper ocean has barely increased and the rate of sea level rise slowed,...they put forward a climate model showing that decade-long pauses in temperature rise, and its attendant missing energy, could arise by the heat sinking into the deep, frigid ocean waters, more than 2,000 feet down.*"

James Hansen, climate scientist: "*All the climate models, compared to the Argo data and a tracer study soon to be released by several NASA peers, exaggerate how efficiently the ocean mixes heat into its recesses....that climate models have been overestimating the amount of energy in the climate,...“Less efficient mixing, other things being equal, would mean that there is less warming 'in the pipeline,'” ....it also implies that the negative aerosol forcing is probably larger than most models assumed.*"

Judith Lean, climate scientist: "*The answer to the hiatus, according to Judith Lean, is all in the stars. Or rather, one star...Climate models failed to reflect the sun's cyclical influence on the climate and “that has led to a sense that the sun isn't a player,” Lean said. “And that they have to absolutely prove that it's not a player.” According to Lean, the combination of multiple La Niñas and the solar minimum, bottoming out for an unusually extended time in 2008 from its peak in 2001, are all that's needed to cancel out the increased warming from rising greenhouse gases.*"

John Daniel, *climate scientist:* "*We make a mistake, anytime the temperature goes up, you imply this is due to global warming,*" *he said.* "*If you make a big deal about every time it goes up, it seems like you should make a big deal about every time it goes down.*"

While there is evidence that even those pushing human made global warming really have a difficult time finding facts that support that idea. They keep searching and some modify data, but most seem to be in love with the idea of cleaning the earth and protecting our environment. They want to protect the Earth from man as we MUST be the reason for problems. Let's continue looking at their data.

Boiling Ocean "Mistake"

James Hansen was at it again. Look closely at the graph below this is what "boiling" oceans look like after some 1.3 trillion tons of CO_2 emissions poured into the atmosphere since 1850. The huge increase in COS levels shown as a dotted line had absolutely NO effect on the Oceans that have stayed the same temperature. James Hansen's belief of CO_2 caused global warming is not supported by the tropic's data in the least nor is his crazy prediction of boiling oceans. You can keep driving you gasoline car and the earth will not even know it.

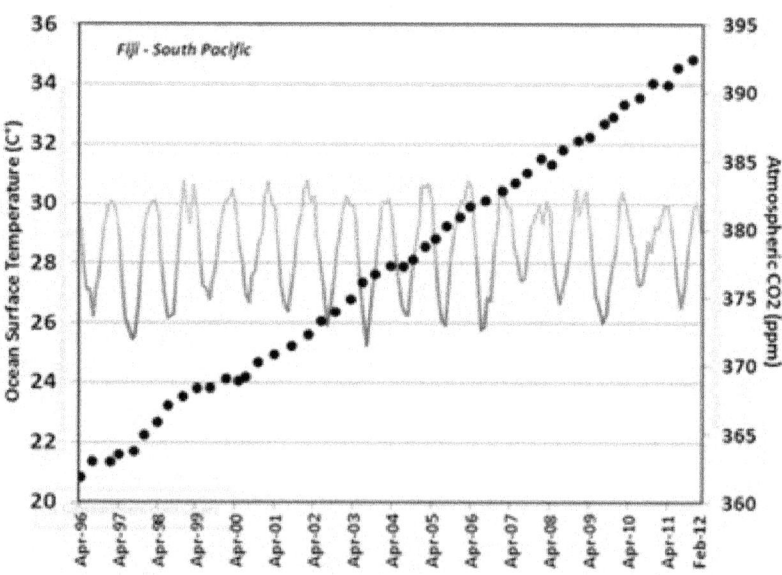

Arctic Warming

We are told that the biggest sign of "climate change" is the rapidly warming Arctic. It is even called the world's thermometer, proof that global warming cannot have stopped. Certainly, the evidence of this from the GISS satellite is persuasive for the Arctic, but not so convincing for Antarctica. The charts are of the average temperature between 1951 and 1980 compared to the average temperature in 2014. The Arctic is getting warmer just as the Antarctic is getting solder. Nobody is showing you the bottom of the Earth. Maybe we should look farther.

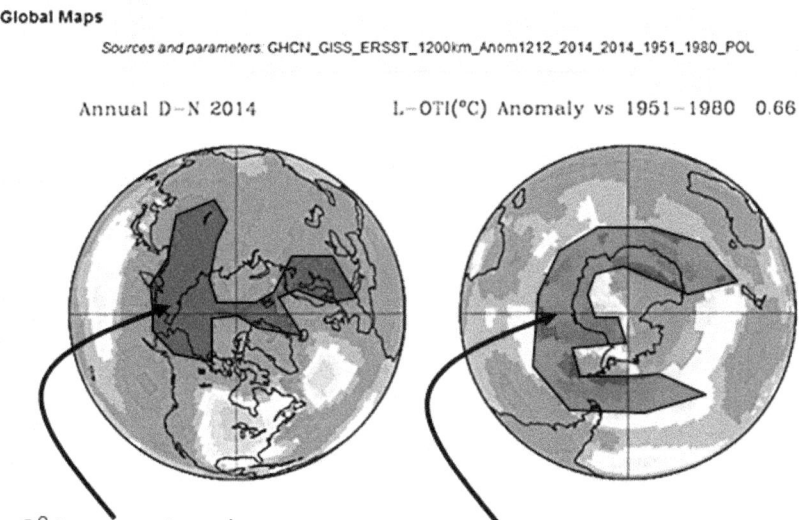

GISS Surface Temperature Analysis

Global Maps

Sources and parameters: GHCN_GISS_ERSST_1200km_Anom1212_2014_2014_1951_1980_POL

Annual D–N 2014   L–OTI(°C) Anomaly vs 1951–1980 0.66

3^0 temperature Increase 3^0 temperature Decrease

Substantial Arctic Change is NORMAL

It is well established that the Arctic warmed up rapidly during the **1930's and 40's**, before temperatures plunged in the 1960's and 70's. James Hansen knew this as shown in the graph below from his 1987 paper *"Global Trends of Measured Surface Air Temperature"*. Coming out of the last mini-Ice Age around 1880, the temperature rose to some nominal level and dropped only to rise again by 1940. For the next 25 years the temperature dropped again by about 1.5^0 to what the alarmists use as the "NORMAL" Arctic temperature. Notice the temperature between 64 and 90 degrees latitude [Near the Arctic Circle]. Also notice the temperature near the Equator had a very slight dip in the Ice Age and a slight increase to 1980.

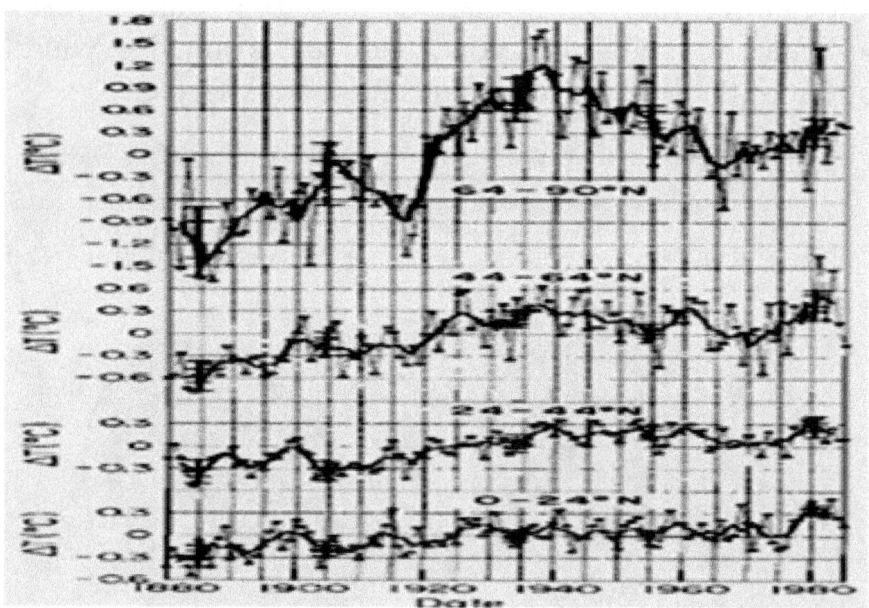

The loss of Ice in the Arctic is a "correction" rather than a catastrophe.

Greenland Ice is Returning

In **Greenland**, Tavi Murray (Swansea University) reported that two of the major glaciers, Helheim and

Kangerdlugssuag, had slowed significantly by 2006, and reported that in 2007-08 that there has been a 'synchronous switch-off' of speed-up of the 14 largest outlet glaciers in southeast Greenland … not a 'runaway acceleration' that has erroneously been reported.

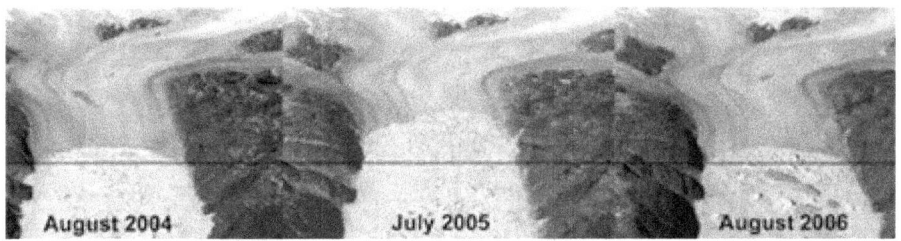

This shows the Helheim Glacier's flow to the sea sped up in 2005, as evidenced by the 5-kilometer retreat of its leading edge, but by 2006 it had slowed back down showing the minor temperature had little effect even in the Northern hemisphere.

Expanding Antarctic Ice

Most of the Antarctic <u>ICE is expanding, not melting</u>. This is contrary to widespread public belief. The Scientific Committee on Antarctic Research of Antarctic Treaty nations reported recently in Washington that, *'cooling at the South Pole was significant in recent decades'*. The eastern part of Antarctica is four times the size of west Antarctica, and <u>ice losses in west Antarctica</u> over the past 30 years have been <u>more than offset by increases in the eastern part</u> of the Ross Sea region. Ice core drilling in the fast ice off Australia's Davis Station in East Antarctica by the Antarctic Climate and Ecosystems Co-Operative Research Center shows that last year, the ice had a <u>maximum thickness of over 6 feet is its densest in 10 years.</u> The <u>average thickness of the ice at Davis since the 1950s is about 5 feet.</u>

Antarctic Sea Ice

It is reported that <u>Antarctic sea ice has *increased* 4.7% since 1980. Over 90% of the World's glaciers are in the Antarctic and they are growing.</u> The following image [left] shows the 2009 Ice extent compared with the average sea ice distance [Thin line]. This is 90% of the World's ice and 70% of the World's fresh water. Sea ice around Antarctica varies from about 8 million square miles in September or October to about 1 million square miles in January. The following image [right] is from NASA is of sea ice maximum image from the Nimbus 7 Scanning Multichannel Multiwave

114

Radiometer (1978 to 1987) showing a similar increase. Yes the area just below South America is showing a slight reduction in Ice, but the rest is getting more ice.

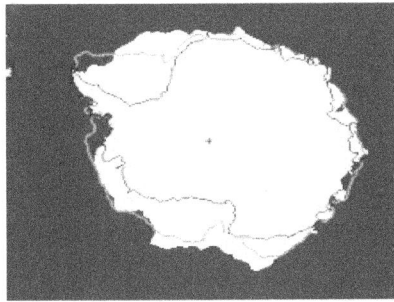 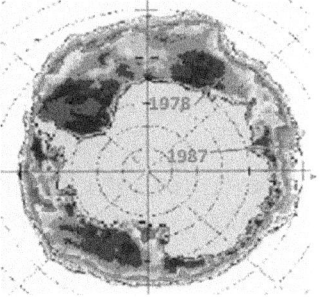

The next image [left] is a NASA composite map of Antarctica showing areas of greatest warming as the darker region. The rest is colder. The Wilkins Ice Shelf lies off the peninsula in the top left corner. Antarctic is approximately 14,000,000 square kilometers and about 8,000,000 square kilometers of shelf ice melts and reforms each year. By picking the small area where the average temperature is getting slightly warmer for your FAKE GRAPH] it looks like Antarctica is losing ice, the oceans are getting deeper and the world is in trouble. The thing to know is that the data [while it can be considered correct, is a total lie.] The graph to the right shows the slight increase in average Ice levels in the southern hemisphere. No one is showing us that graph.

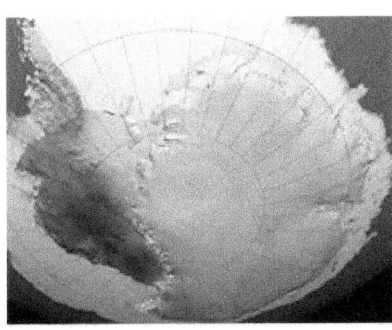 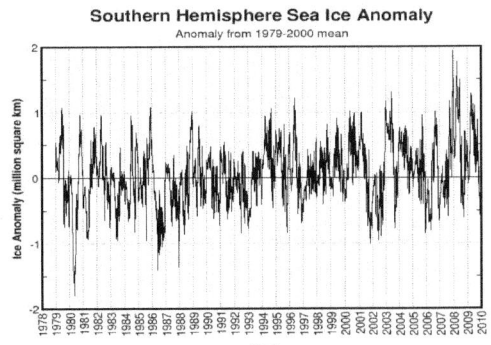

CO2 Villain Mistake

Here we find that CO_2 in the atmosphere went up slowly from 1870 to 1970 (290 parts per million, ppm to 320ppm). It then, by fantastic fiction, the CO_2 level rose a further 70ppm to about 390ppm in 2010. This would be the vanguard of disaster. Everyone would melt down. The graph below shows the CO_2 content captured in the Ice Core samples of Antacrctica.

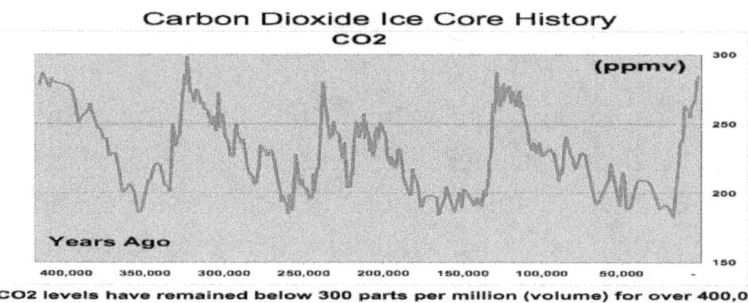

CO2 levels have remained below 300 parts per million (volume) for over 400,000 years

The following graph shows the faked increase sice 1600 or 1700. While there was a slight increase until 1996, No published Ice Core samples have been made since. All data since them comes from aeresol CO_2 in Hawaii.

Let me show you how it was done. The first graph below show the transition between CO_2 measurements in Antarctica ending in 1996 and Aeresol CO_2 measurements from Hawaii and from satelite imagry that started about 1960 and is shown with directed data up until 2008. Notice there is a 33% difference in the absorption levels of CO_2 in antarctica and the aeresol figures. Knowing the abosorption level, simply reduce the slope of Aeresol CO_2 by a 1/3 and what you have is shown in the middle graph showing a very slight CO_2 within "NORMAL" levels.. Instead of using logic, NOAA used trickery and misdirection to make the last graph that has been scaring everyone, including climatologists wanting there to be global warming to allow them to write papers and gain notoriety.

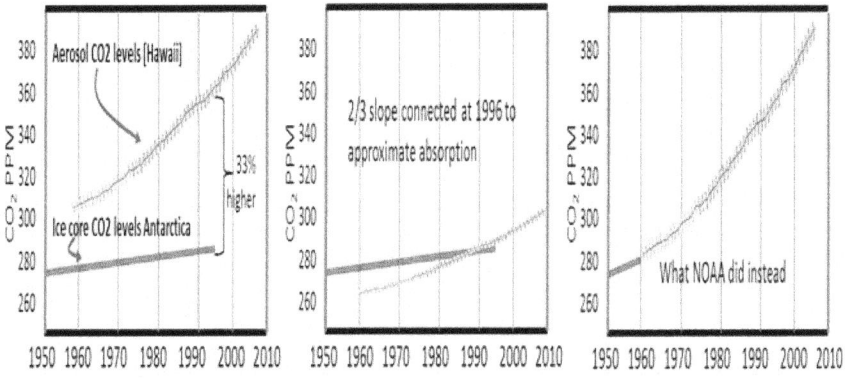

By putting the graphs together to make it look bad, they simply attached the Hawaiian levels onto the Ice Core data as if 100% of the CO_2 in the Hawaiian air gets to Antarctica and is nott recombined before being buried in the Ice.

Stupid, Stupid, Stuipid

The problem is we aren't talking about stupid people we are talking about dangerously bad ones. Ignore the name on the graph as it is another lie.

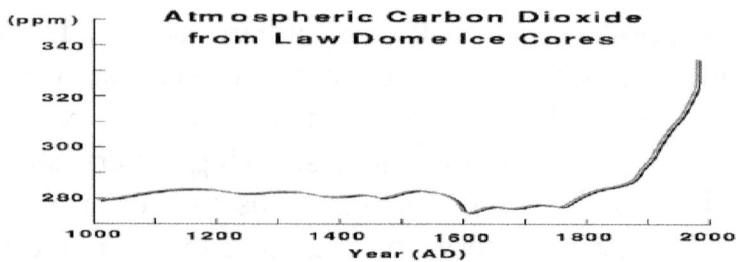

They did the same thing with NO_2 levels and CH_4 levels and they produced terror to reap benefit.

Expanded Data

Have you ever noticed that these "sicientists" never show expanded views of the Ice Core data? There is a similpe reason for this. While they need the very ancient data from Ice Coring to show how our temeprature is now going out of control, they don't want you to know about inconsistencies between satellatie imaging and Ice core concenrtration graphing. I thought I would just show you a couple so you can appreciate how devious these people are and so desperate to convince you of their made up disaster. On the left is the data from Greenland Ice core temperature mapping and the satellaite temperature map is shown below it for the same general time. To the left is a similar interaction with data from Antarctica. Please see that there is lost data suppressed by those wanting to show a particular evil wen it doesn't exist. The high temperatures from the 1940s is completely missing from the NOAA 2015 data well after the falsehoods I already brought out earlierr were uncovered and proven. While the Antarctic data looks more consistent, there till are errors as the downward slope of the Ice core is converted to an upward slope from the NOAA data.

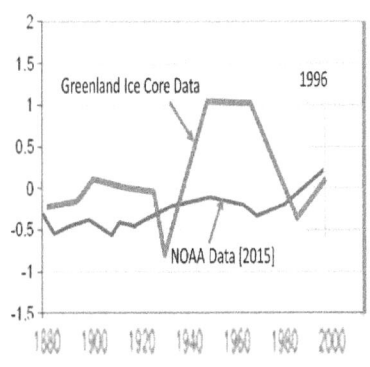 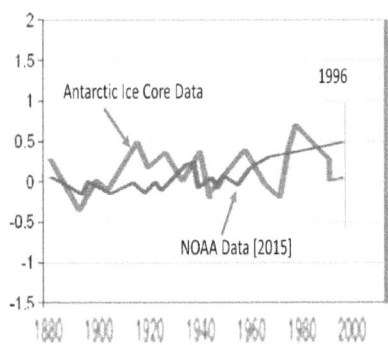

IPCC Data Tell Us The Treachery

Luckily for you, you don't need to take my word for anything. The IPCC told on themselves in a paper presented in 2009 that tried to confirm massive increases in CO2, but what they ended up showing was TREACHERY. Here are some snippets. I have not put them together just to make it look bad, but I didn't think you would want to go through all the mess so I boiled it down a little. The IPCC acknowledged CO_2 has something called a short residence time, stating:

> *"The turnover time of CO_2 in the atmosphere is about 4 years. This means that on average it takes only a few years before a CO_2 molecule in the atmosphere **is taken up by plants or dissolved in the ocean**.*

As you read through this remember the Hawaiian Aerosol CO2 detection is the thing that is causing the entire ruckus. The data taken included substantial amounts of CO_2 that would be absorbed into plants and never get into the Antarctic Ice.

> *"The CO_2 response function used in this report is based on the revised version of the Bern carbon-cycle model used in Chapter 10 of this report. --About 50% of a CO_2 increase*

119

will be removed from the atmosphere within 30 years and a further 30% will be removed within a few centuries. The remaining 20% may stay in the atmosphere for many thousands of years".

Let me first say that the 30% that won't mix with Ice for centuries should have been used in the amalgamated data. Let me just say if 20% CAN'T ever get into the Ice Core HOW IN THE WORLD could some crackpot just add it in? I could tell you why, but I would have to spit a lot. Let's continue.

"A quasi-equilibrium amount of CO_2 is expected to be retained in the atmosphere by the end of the millennium that is surprisingly large: typically ~40% of the peak concentration enhancement over preindustrial values (~280ppmv).

Here they are trying to say, ignore what I just told you --- never mind, I have no idea what they are saying it is complete gibberish, but we have to continue.

"If the partial pressure of CO_2 varies and the hydrogen ion concentration was kept constant, the relative changes would be the same in the sea as in the atmosphere. As the total amount of CO_2 in the sea is about 50 times that in the air then practically all excess CO_2 delivered to the atmosphere would be taken up by the sea when equilibrium has been established".

Oops, I don't think they were trying to say, the Aerosol numbers had no effect on the Ice Core samples. Maybe I don't understand their English.

Estimates of past carbon dioxide concentrations derived from ice cores drilled at Vostok, Antarctica and Siple Station, Greenland are combined with the modern instrumental record from Mauna Loa Observatory to show the relationship between atmospheric CO_2 changes associated with ice ages and the modern increase in CO_2 associated with human activities. Natural control of atmospheric CO_2 ended at the time of the Industrial revolution, when humans began burning fossil carbon fuels, manufacturing cement, and removing forests at an increasing rate.

This makes you wonder if they read their report before telling the world that they were going to ignore it and everyone else should as well and they should somehow understand that all the CO2 in Hawaii gets to the Ice cores somehow.

CO$_2$ Anomaly

Something else doesn't add up. If the massive increase in CO_2 charts were in anyway correct and CO_2 was supposed to be the harbinger of global disaster. While CO_2 was increasing slowly over that total period, global mean temperatures actually went <u>down</u> *twice* in that period; first between 1870 and 1915 and then again from 1940 to 1970. To make things even weirder, temperatures actually went up rapidly in the 1915 to 1940 period when the change in CO_2 was almost flat. Luckily for those with stock in windmills, there was **one** period (1970-2010) that showed similarity as CO_2 and temperature both increased. With 3 out of 4 periods **not** matching, it is inappropriate to think there is a correspondence between CO_2 and temperature? These trends are shown in the following figure. We can only conclude that there is a poor correlation between

121

CO_2 and earth's temperature, but Al Gore still got a Peace Prize for ignoring most of the data. Later we will look at something that does track, but it is not controlled by forcing people to drive electric cars. The following chart shows wide variations in Arctic temperatures [the tope line] with almost no change in CO_2 [bottom line] until a couple hundred years ago when CO_2 levels began rising and the Temperature stayed almost stationary. No one in their right minds can tie CO_2 to temperature fluctuations.

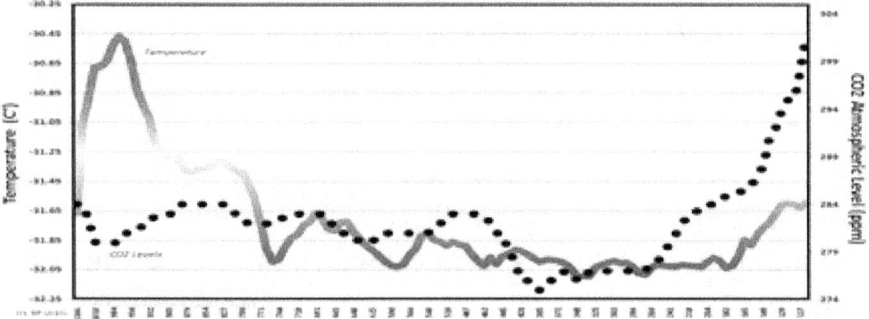

Today we have an even larger differential as shown next. The average temperatures have settled and are slightly decreasing while the CO_2 levels are skyrocketing according to Greenland data 1997m thru 2012 as shown below.

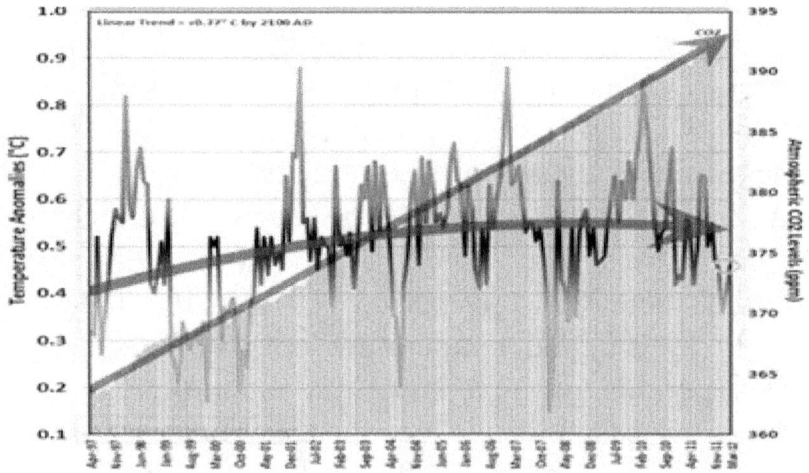

The empirical evidence is so overwhelming that even the vast majority of alarmist climate scientists (over 97%) agree that the predicted "accelerating" global warming has been non-existent over the last 15 years.

CO_2 Absorption

Besides all that calculation of how long CO_2 stays in the air, it can't get in the air easily. Those suggesting CO_2 is a major greenhouse gas, seemed to have never even looked at an atmospheric absorption map as shown next. Water is by far the most absorbed large molecule gas that can cause horror or blessing. It is very difficult to absorb CO_2 into air. Even in areas where CO_2 can absorb, water is already so absorbed that it still has an inability to be introduced in our atmosphere. Here is what we can suppose about the large increase in CO_2. There has been a tiny reduction in water vapor which shows up as a much higher CO_2 percentage. No matter what, we must understand that water vapor controls our atmospheric temperature----not CO_2. CO2 has only 3 small wavelengths of absorption while water is absorbed in huge amounts.

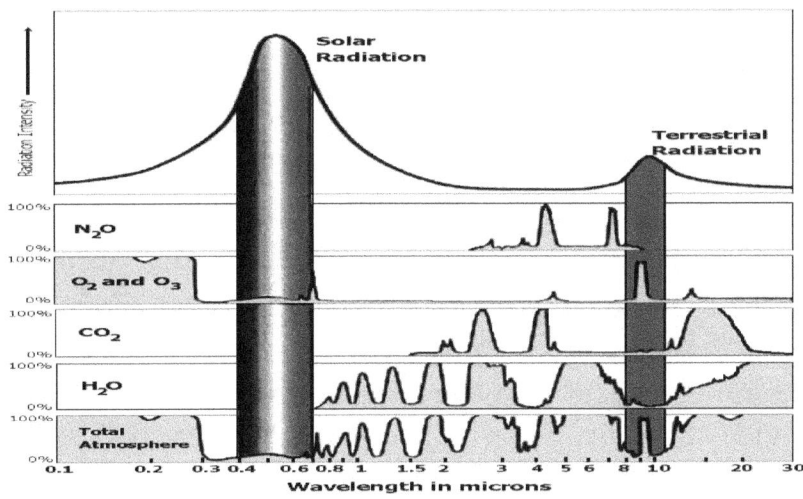

Some have speculated that when the CO_2 absorption gets to 560 ppm there will be a substantial temperature rise that could be devastating. [Oops! I laughed a little writing that one.]   A massive rise in CO_2 already noted has driven our temperature to the unbelievable increase of about 0.3 degrees making a further increase to 560ppm form the current 390 have a total increase of less than 0.5 degrees. We simply are not affected by CO_2 in the atmosphere at all. 560ppm (the dreaded doubling), temperatures should rise by *another* 0.2   to   0.5°C *ONLY.*   IPCC   [NATOs   Global Warming gurus] estimate of 2.0 to 6.0°C, and this totally unfounded and without scientific merit.

Not much CO2 in the First Place

If CO_2 in the atmosphere has risen from about 290 to 390ppm in just over 100 years, which has been recognized, and if only 5% of all the greenhouse gases are man-made, then we can conclude that only 5% of the extra 100ppm could have been caused by mankind. This is only 5ppm over 100 years so we can say the following:

*The other change is due to the 95ppm from **naturally produced CO_2**.*

Added to this, IPCC claimed that we will get 3-5^{0}C more warming in the next 50 or so years. Mankind would only contribute 5 to 10% of that. That is only 0.2 to 0.5^{0}C, even if we keep using coal.   Just think about it man-used CO_2 makes up only 5% of the 5% of the 1% of the atmosphere or 0.0025%. The following chart represents one view that is much less than 5% for man uses and remember CO_2 makes up the tiniest fraction of the atmospheric Greenhouse gases, so divide that tiny piece of the pie down much farther and see if it can have ANY effect on weather.

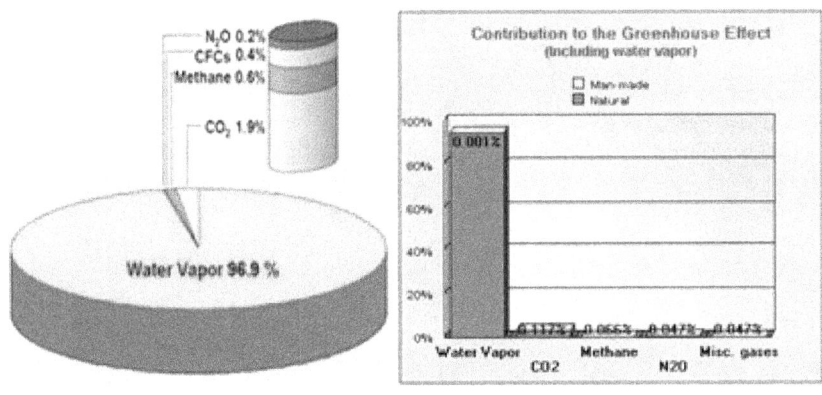

CO2 and Temperature

Let's hypothetically say CO2 is changing our temperature. The charts below show both CO2 levels and temperature captured in the Ice Cores from Greenland and Antarctica. The one on the left is from Antarctica over the last 50 thousand years. The thin line that begins below the erratic temperature curve shows something interesting. CO2 doesn't change until after temperature changes as temperature controls CO2 rather than the other way around. The second graph is from Greenland in case we didn't see things right, we see that Temperature, the erratic line changes well before CO2.

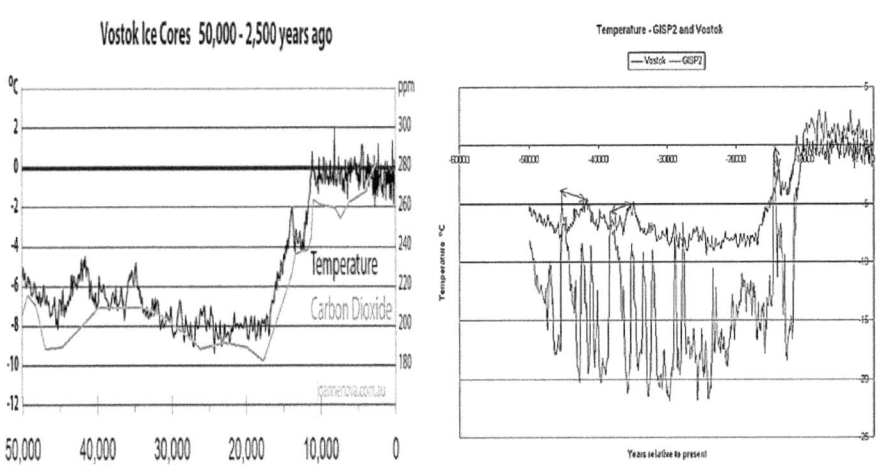

Methane

Methane (CH_4) is produced through both natural and human activities. For example, natural wetlands, agricultural activities, and fossil fuel extraction and transport all emit CH_4. Methane is more abundant in Earth's atmosphere now than at any time in at least the past 800,000 years. We are told, due to human activities, CH_4 concentrations increased sharply during most of the 20th century and are now more than two-and-a-half times pre-industrial levels. In recent decades, the rate of increase has slowed considerably. As I showed earlier, the reason Methane looks like it was running away was the mixing of Ice core data and the aerosol data from Hawaii. As I showed before the one you see is below. Simply chop off the stuff after 1990 it has no place on that table. This is especially true because of where we find it.

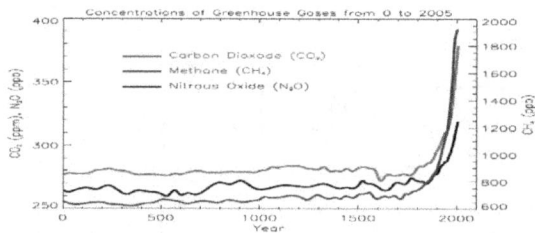

Where is Methane Produced?

We know where methane is from satellite data shown below. The darker the color the more intense the methane production and aeresol levels. Methane is made on the Equator and by the time it gets to Antarctica to combine with Ice Core data it is so bogus that it just made be throw-up a little.

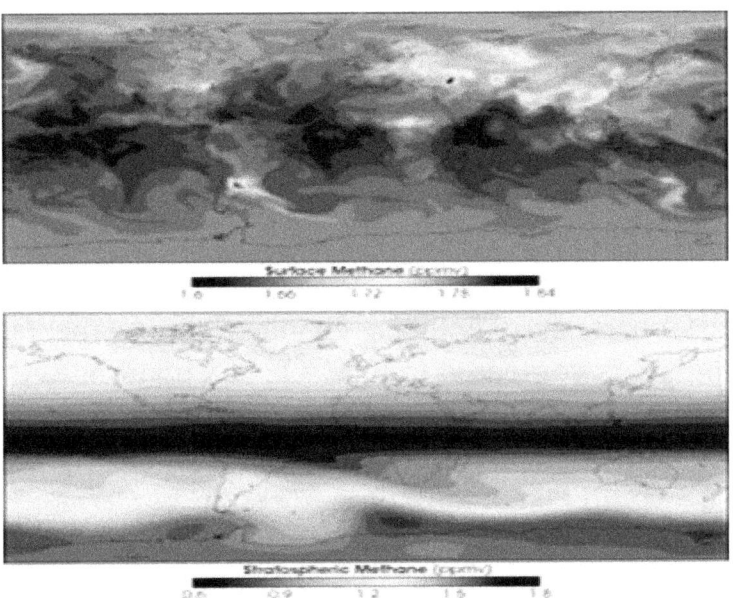

The top graphic shows Computer models showing the amount of methane (parts per million by volume) at the surface (top) and in the stratosphere (bottom). Along the equator the concentration is huge and the polar regions get almost no methane. The idea that the methane levels would show up as this massive increase just because Hawaii showed a larger concentration of methane is just plane stupid. If not stupid, it is criminal to make it look like there has been this huge increase mapped with Ice.

Arctic Methane

EDML is Antarctica or (EPICA) Dronning Maud Land site and EDC is Antarctica (EPICA) Dome C site. Please notice massive CH4 fluctuations 128 thousand years ago. These are real fluctuations rather than those demonstrated today.

127

We can imagine that if the CH4 fluctuated so much in Antarctica, Hawaii during that time would have been really cooking and there was no Greenhouse catastrophe.

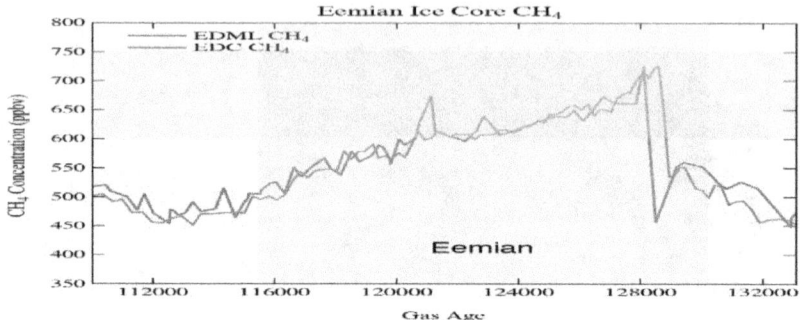

The last chart showmethane changes across the last 180 thousand years compared to the CO2 fluctuations. As I said before, Aeresol variations of these gases in Hawaaii would have been substantially more dramatic and slightly older are the gases would take a long time to settle.

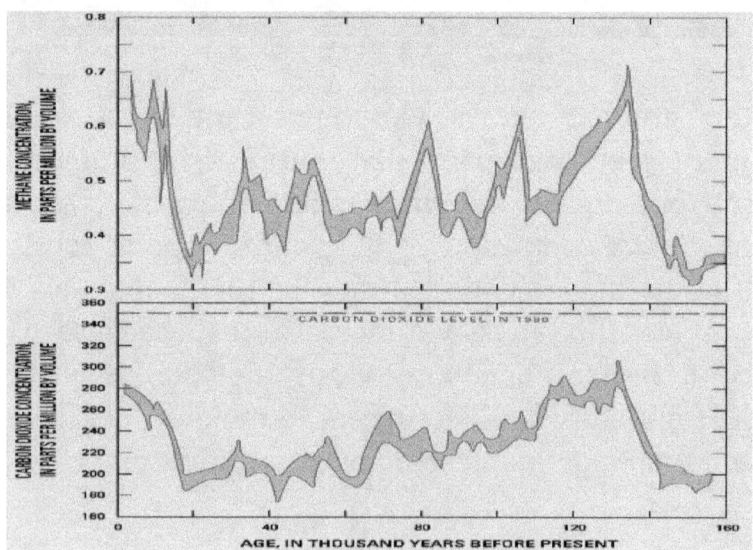

Fossil Fuel Conversion

The big focus has been on converting what is called fossil fuels into bio-fuels made from grain. The thought is that there is a tremendous amount of methane and CO2 manufactured by combustion engines. This is the reason for dumping the Hawaiian data on top of the Ice core data. The problem is that neither of these things makes a significant temperature problem and H_2O is the mastermind of green gas. Hopefully, by now, you are at least questioning the data.

Bio Fuels

Bio-fuels from grain are not helping the situation. Use of "grain based fuels" will greatly increase food prices and roughly 30 million people are expected to be severely deprived. The USA will use up to 30% of the annual corn crop for alcohol production for vehicles alone. Ethanol production requires energy too to make it economically. The actual cost/gallon is much the same as other liquid fuels, but the miles per gallon consumed by vehicles are much higher than gasoline. One estimate is that one tank full of ethanol for an SUV is obtained from enough corn to feed one African for a year. Of course it isn't working so worldwide ethanol plant subsidies in 2008 alone totaled more than $15 billion.

Dinosaur Flatulent Methane

The federal government is spending millions of dollars researching one of the largest producers of ozone destroying methane—the cow butt. This whole concept of cattle flatulence destroying our ozone makes you think about dinosaurs. As much as cattle expel methane, dinosaurs must have been even worse. They must have been awful to travel behind. From the high levels of the almost impossibly fossilized remains and the huge quantities of oil left behind we can be assured that there were more dinosaurs on this planet than we could possibly imagine. From them, the methane levels in the very ancient past surely would have been enough to de-ozone-ate the atmosphere and the old earth should have been consumed in a ball of fire like many scientist indicate we are heading for if we don't stop up the pesky cattle butts. I think we should leave cows alone. It didn't happen during the ancient smelly time and the likelihood of it happening today is slim to none. We are probably going the other way.

Little Effect

We could go on and on as new "Greenhouse industries" to eliminate water production would make millionaires of the proponents that saw this horror as water heats our earth to alarming levels. Instead let's attack Volcanoes.

Do Volcanos Do Anything?

For completeness, let's look at these things. If you were looking for a CO_2 emitter you would be on the right track. Volcanoes are active around the world and continue to emit carbon dioxide as they did in the past at the rate of 130 to 230 million tons of carbon dioxide per year but burning what are called fossil fuels, releases about 26 billion tons of carbon dioxide, into the atmosphere every year so we can disregard volcanoes as well.

Sulphates

Sulphates have been measured in samples from the Greenland ice cores. A general description of what was found is shown below from 1880 until about 2000.

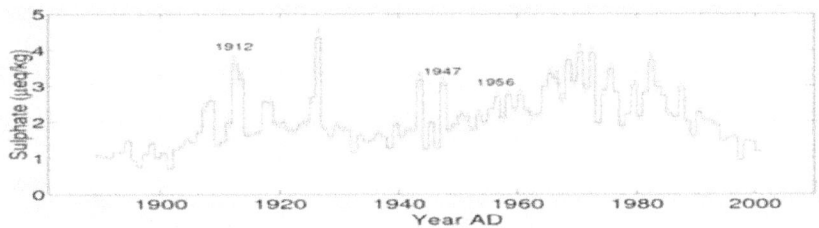

Notice that volcanoes push out Sulphates. The volcanic imprint is seen in the early part of the record: Katmai

131

(1912), Hekla (1947) and Bezymianny (1956) all caused elevated levels of sulphate, but during the 1940s through the 1970s, the chart shows that something was still lingering in the air and finally began to be deposited at elevated levels. We can believe most was from the remains of the predecessor volcanic action. In this period, sulphate pollution caused acid rain and severe damage to trees in some areas. Some believed that sulphates should all be falling to the ground overnight and the bubble was from massive fossil fuel automobiles of the 60s. The bubble dissipated between 1960 and 2000. Some claim the reduction was the elimination of Sulfates in automobile emissions. I believe this is a good thing as I like less acid rain, but the timing suggests there were other dissipaters like fewer eruptions at this time.

Nitrous Oxide

Nitrous oxide is produced through natural and human activities, mainly through agricultural activities and natural biological processes. Fuel burning and some other processes also create N_2O. Like the other claims, concentrations of N_2O have risen approximately 20% since the start of the Industrial Revolution, with a relatively rapid increase toward the end of the 20th century. Here is what you were told. Overall, N_2O concentrations have increased more rapidly during the past century than at any time in the past 22,000 years. The big problem with that statement is that we do not know what the airborne Nitrates were back then. All we have are the levels that got into the ground and were frozen. The slight increase from 1880 until 2000 is shown below left.

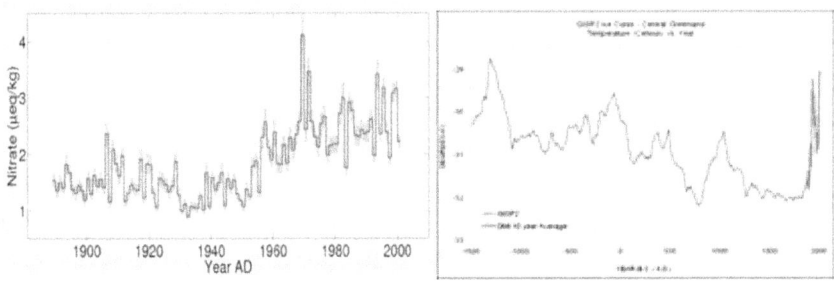

Lead Pollution

The Chart above right is one of those where they used the Ice Core data for everything before 1990 and then switched to Aerosol data making it look like wild fluctuations over recent times. Please don't look at the last part. All these values are very tiny and have been blown up to make them look significant, but there are only tiny amounts in the air.

Black Carbon

Black carbon is sort of a new worry. It is a solid particle or aerosol, not a gas, but some tie it also to warming of the atmosphere. It is true that black carbon particles can directly absorb incoming and reflected sunlight in addition to absorbing infrared radiation very slightly, but certainly not as good as gases that can actually bond with other air particles. Here is the worry. This black carbon can be deposited on snow and ice, darkening the surface and thereby increasing the snow's absorption of sunlight and accelerating melt. If you do see Ice getting black, please scrape it a little so the Arctic won't melt. This will not slow down the periodic melt cycles and it is nothing to worry about.

The Sun

All those greenhouse gas things are the real reason for temperature fluctuations. The real culprit is the sun. All of he heat in our atmosphere is solar heat. If the sun gets brighter and burns away clouds, things happen and when the sun generates cosmic and X-Rays that hit the earth other things happen, Instead of just shining, the sun blast energy out in spurts.

Solar Flares

If we look at the correlation between temperature and total solar irradiance we see a much better relationship and we can begin to understand CO_2 is not the main player. In fact; the tiny CO2 gas levels could not possibly affect weather no matter how many gas guzzling cars there were! It is the activity of the sun (sun spots, solar flares, modification of other galactic cosmic radiation from outer space, the effects of solar wind, and magnetic flux), that affects the radiation arriving on earth.

Here is a big one. The sun moderates cloud cover! Approximately 1% of the atmosphere is greenhouse gas and 90-95% of that is water. CO_2 is about 0.05% of the atmosphere. But only 5% of that 0.05% is man-made!

Sun Spots

The sun is 1.3 million times larger than the earth; when its temperature changes, our temperature changes. Well, think about it. Every year, the temperatures rise and fall with spring, summer, fall, and winter. A year is simply a 365-day cycle. Every day, the temperatures rise and fall with daytime and nighttime. We can neither change them nor stop them any more than we can stop the Earth's rotation. The temperatures fluctuate based on these cycles. So clearly, the Earth's temperatures rise and fall based on its exposure to something we call the sun. There are larger cycles of the sun called "solar cycles." The following graph shows the cyclic nature of sunspots. Notice that as the temperature increases, the amplitudinal difference of the solar flare cycles increases.

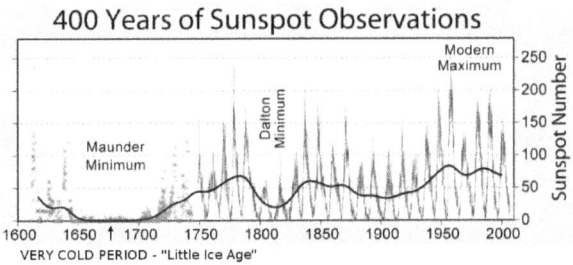

From the next graph we can see that from 1978 the temperature is getting slightly warmer in parts of Greenland, so what did we do differently before 1978. The graph shows that there has been <u>no appreciable slowdown in the increase of CO_2 in the atmosphere</u>, but there was a fairly significant reduction in the temperature between 1940 and 1978.

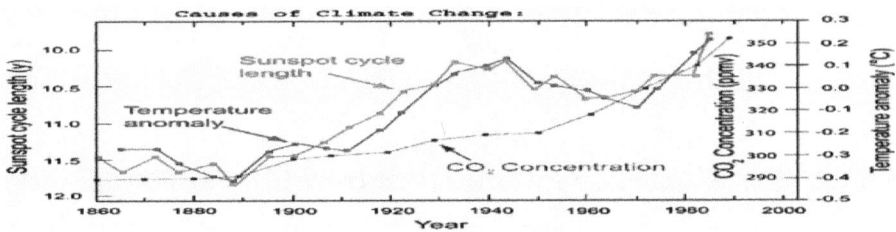

The 400 year graph following shows a potential culprit. Sunspot activity is cyclic and when there are more spots, the temperature rises. Some have suggested that if we could simply shield the earth from sunspots, our global warming issues would evaporate. There is a consistent similarity between sunspot activities. While CO_2 levels simply were rising since the 1800s, sunspot cycles go along with tmeperature rise and fall.

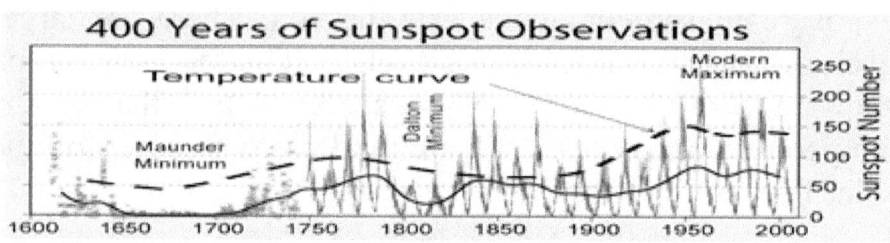

The sun gets hotter and times when it cools off as measured by "sunspots." If the climatologists were paying attention to these "solar cycles" years ago, they could have told you that the Earth would get warmer during the 1990s, and then it would begin to cool just like it has. The next graphs may help us understand some of what is going on as the magnetic field of the sun has been reducing over recent years as our earth has become a little warmer.

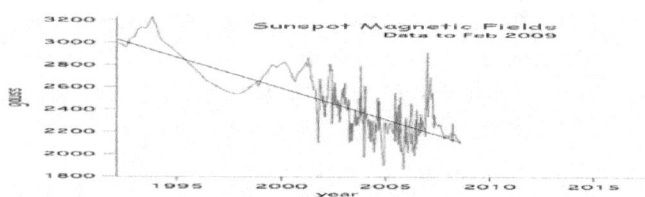

Solar Brightness

Besides the magnetic field fluctuations, changes in the brightness of the Sun can influence the climate from decade to decade, but an increase in solar output by itself only changes by about 0.1% between hottest and coldest times and the current direction is just now starting to try to increase the average temperature as shown next left.

The sun's irradiance has an affect on our temperature that should be recognized as shown next right. It shows that while CO2 levels are changing at a constant rate, the intensity of the sun is cyclic just like the top two Temperature lines reveal. That brings us to Cosmic Rays.

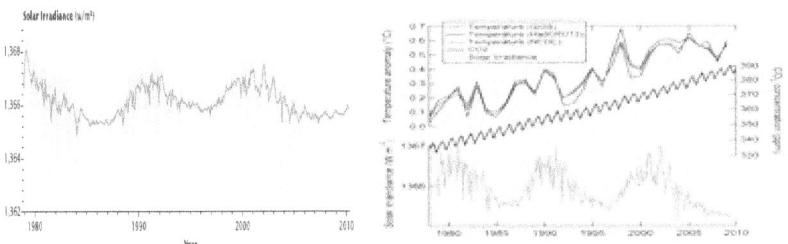

Cosmic Ray Temperature modification

As shown next we can say global warming cycles are seem to also be associated with Cosmic Ray cycles, X-Ray cycles and Sunspots. Certainly someone would be showing this stuff to you so you would see the sun's control over your temperature, but somehow, they forgot to tell you.

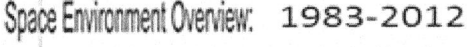

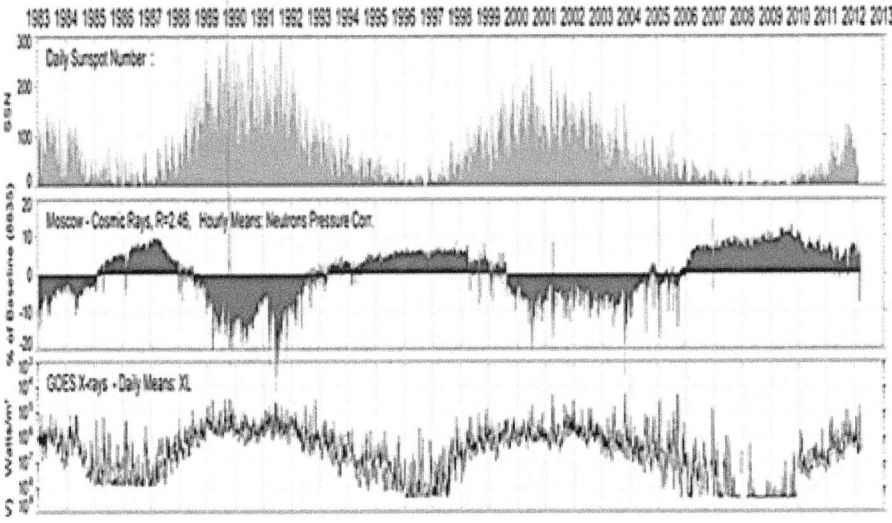

Climate Control

Attempts to correlate solar activity with global temperature have been going on for some time now as you would expect. Measurements from the SORCE's *Spectral Irradiance Monitor*s show that solar UV variability produces colder winters in the US and southern Europe and warmer winters in Canada and northern Europe during solar minima. Here is what we are finding.

Cosmic Clouds

Solar wind-mediated galactic <u>cosmic ray</u> changes, seems to affect cloud cover which would change the thermal characteristics. Much of the solar effects have been clouded by the evils of NOAA trying to keep their favorite stocks going up, but I think you can see climate is almost totally controlled outside our atmosphere. The cosmic ray change over the cycle changes clouds. Let me be direct here!

All environmental scientists know this and they also know that the most significant changes are in the polar regions.

Therefore, the poles show more thermal variance than other parts of the world. Get rid of Cosmic Rays and the Poles will stay more regulated, but making sure cows have less flatulence will not save the penguins

What Causes Solar Cycles

Just so you have a little background solar cycles are just like tides. Like out moon, Jupiter causes variations on the solar surface and it is attributed as one of the major cause for cycle sunspot and other solar activity. If we could get rid of Jupiter, we could make our temperature cycles reduce, but eliminating coal burning just removes someone's livelihood. After seeing how clouds, might be something to look at, let's do it.

Clouds and Water

Between 1990 and 2000, a 5% decrease in cloudiness *increased* the total earth surface radiative flux by 6 Watts per square meter. The entire CO_2 forcing as estimated by the UN's IPCC climate panel was just 1.6 Watts per square meter over the past 250 years. While CO_2 can cause warming, its role is tiny compared with natural influences, and much, much lower than the UN's IPCC estimates. Water, water vapor, and clouds are dominant, but here is the rub!

Only CO_2 can be taxed!

Recently the Weather Channel announced that at any one time there were 20 **trillion** gallons of water in the atmosphere over the States at any one time. This equates to 75,000 million tons of water, or 10,000 tons of water in the air for every square kilometer. I am certain the members of our governments taking part in this travesty would greatly desire to tax all that green gas. Just think about we would have to control our waste so that none of the water gets in the air and taking a shower longer than 2 minutes would be severely taxed. The following graph may tell us some important information. We can see that the higher altitude Stratosphere [the line that starts highest and ends lowest] is much more dynamic and getting cooler while the lower Troposphere is fairly steady and getting slightly warmer showing. Ta-da!!!!! The lower air is different than the Stratosphere line because of-------CLOUDS controlled by cosmic rays.

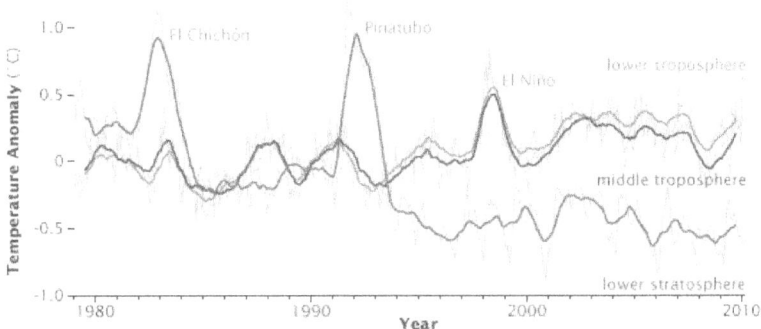

Most Dangerous Greenhouse Gas

The IPCC can't see this simply because they don't consider the main greenhouse gas as water vapor. If _95% of the greenhouse gas is water_ and it is; we must attribute clouds as one of the major causes for temperature increase. Let me give you an example. With CO_2 at 390 ppm and water vapor is somewhere between 10,000 and 40,000 ppm and wreaking havoc on our atmosphere. When can we start to be honest and see that if we have been in cooler periods, it is natural to expect that it will get warmer so quickly, there must be a much larger driving force than a trace of CO_2 gas in the atmosphere? The Sun is a major player [man is NOT]. _**Cosmic rays from interstellar space modified by the sun's solar wind**_ plus the sun's and earth's magnetic fields are next largest factors after the sun itself. Following this is CLOUDS. Way below all these is what we call non-water greenhouse gases which play a

141

very minor role. Here is something the NOAA didn't tell you.

If the CO_2 is removed from the atmosphere, water vapor will absorb that band of infrared energy and make just as much heat as if the CO_2 had never been there.

As Cloud cover changes so does the Ozone.

Ozone

Tropospheric ozone [O3] has a short atmospheric lifetime and is a potent greenhouse gas. Chemical reactions create ozone from emissions of nitrogen oxides and water in the presence of sunlight. As the sun gets hotter, we get more Ozone and a cooler sun reduces the Ozone. We like to thing Ozone is great stuff, but in addition to trapping heat, ground-level ozone is a pollutant that can cause respiratory health problems and damage crops and ecosystems.

Ozone Hole

I know you've heard of this stuff and the solar cycle matches the Ozone hole cycles as you would expect, but someone seems to be trying to force fit other things to gain wealth. The image following shows the almost closed hole over Antarctica during the 70s and the larger opening in the 90s [second row]. The last row shows more recent patterns where the opening almost closed again in 2002 and appears to be closing again by 2012. <u>UV light hitting the Earth surface varies by as much as 400%</u> over the solar cycle due to variations in the protective ozone layer. In the stratosphere, ozone is continuously <u>regenerated by the splitting of O_2 molecules by ultraviolet light</u>. During a solar minimum, the decrease in ultraviolet light received from the sun OBVIOUSLY leads to a decrease in the concentration of ozone. This allows increased UVB levels to reach the Earth's surface. I believe you can begin to see this is a chicken and an egg overview. Let's just say that when the Sun emits more UV, the Ozone hole gets larger and during

143

higher activity, it gets smaller. Get rid of the sun and no more OZONE FEAR.

Polar Ozone Depletion – The "Ozone Hole"

Other Global Warming Factor

Planetary warming has been observed on Mars, Jupiter, Pluto, and on Neptune's largest moon Triton during the decades following the peak of the "Solar Grand Maximum" [Its cyclic highest intensity]. Wouldn't you think that Astronomers would come out and say- *there are no humans there!* Maybe some of the emissions from our automobiles are getting to Neptune!!! Sorry! I'll try to not be so accusatory and cynical.

If the Solar Maximum intensity cycle is over and we are going towards a minimum, maybe this is an indicator that the opposite of warming is in the play. Here is what happens. The Earth heats up after a Solar Grand Maximum, lagging a bit after the peak. With a Solar Grand Minimum now on its way, a "global cooling" is on the horizon.

Current research shows that Earth's oceans are now beginning to cool, no matter what you have been told. This is because the temperature cycles in oceans are caused by Solar and Lunar cycles. Instead of confirming these changes, the very groups set up to give us advanced warnings are perpetuating fear and advocating spending billions of dollars on non-solutions for global warming, something that is not even occurring. Although humans contribute to greenhouse gases, the overall effect is a tiny fraction compared to natural causes. To say humans are the cause of global warming; and to also make predictions that global warming is occurring and will continue to increase is simply inaccurate. The "Solar Grand Maximum" has ended

Earth may be headed into a mini Ice Age within a Decade Physicists say sunspot cycle is *'going into hibernation.'*

Closer Look at Temperature

While the Southern hemisphere shows to be getting cooler faster, let's look at what we have in the northern hemisphere. We know it's been getting a little warmer, but is there anything we are missing? As the chart following shows, since the end of the Pleistocene caused a massive increase in temperature allowing us to have much shorter and less cold Ice Ages, There has been a cyclic temperature exchange about every 450 years with the last "Little Ice Age] happening around 1650.

Given this statistic, we will be experiencing the next Ice Age in the next few years and there will not be a huge warming cycle.

This is not ticks of statistics; it is the nature of our planet. Just look at the compressed temperature cycling of the northern hemisphere below.

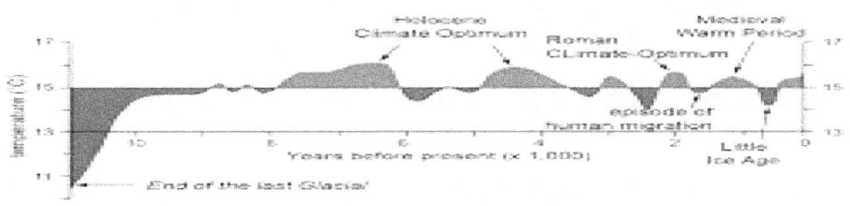

Ignore the Last Ice Age

147

The Global Warming-ites hate it when someone brings up the last Ice Age which lasted from 1650 to about 1700. The well-respected NOAA scientist don't even mention it if they don't have to. Not discussed, mentioned, directed, mapped or characterized, does not mean it didn't happen. The quasi-scientists pushing the return to bicycles and elimination of efficient factories had no effect on this last cold chill and it will have no effect on the next one we cannot stop. Our average temperature rose 2.2^0C from about 1700 to 1735, less than 40 years. That is more than double the effect of the last century blamed on poor old CO_2. Clearly CO_2 did not cause that! I showed you the temperature variance from the Northern hemisphere; let's look at a more dynamic one in the Southern hemisphere. Notice that this one shows a fairly substantial average temperature decline.

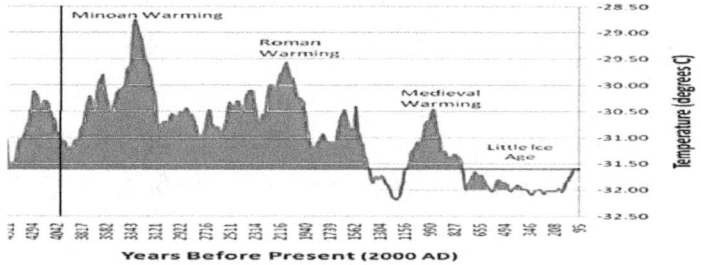

We should look at both graphs together. Both are telling us the same thing. Don't worry about tanning lotion, worry about heating your home.

Sea Surface Temperatures

Here is another alarming bit of data passed on by those with ulterior motives. The Sea surface temperature as measured by NASA satellite which shows a general downward slope, but that isn't all. The average from 1998 till now has been drifting downwards and since 2009 the temperature has been declining continuously.

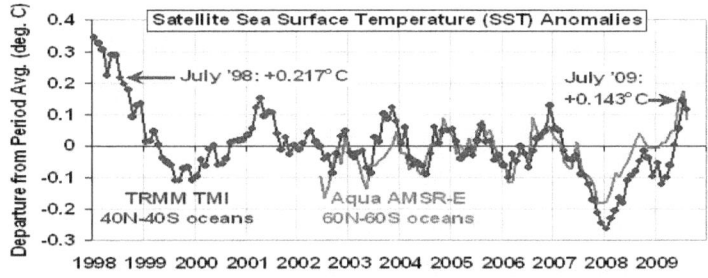

Temperature Trends

If we expand out to the last 4000 years, the Greenland Ice core shows temperatures mostly stayed the same the whole time except for the little dip we started coming out of about 300 years ago. Since that time, the temperature has been trying to get back to NORMAL.

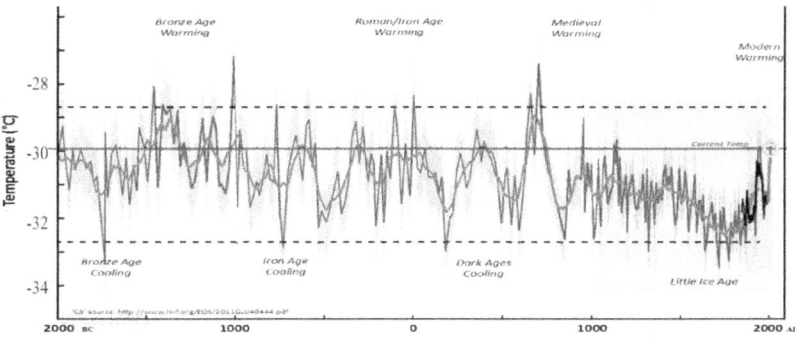

Let's look farther back and see what Greenland's temperature has looked like over the last 10 thousand years as the Earth recovered from a massive shift in its axis. The temperature of Greenland showed the massive shift as temperature skyrocketed to the levels experienced in the current Holocene Age as shown below. Once it stabilized,

149

there were much greater temperature extreme shifts 8 thousand years ago that anything we have experienced in the near term and the Earth did not die.

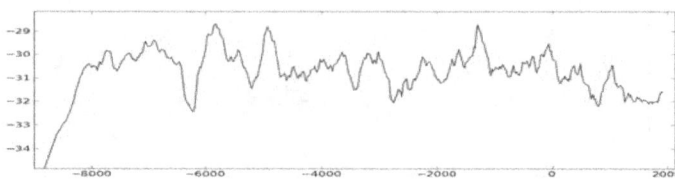

Let me just move back to 11 thousand years ago as the Earth temperature was disrupted by a primary shift before the final one 10 thousand years ago. As shown below. Hopefully you can imagine how horrible it was. Mammoths froze in place in Siberia, piles of animals were twisted together in Alaska, and hundreds of feet of fish dried in the mud at Karoo Africa. With all the shifting the atmosphere would have gone crazy. As the atmosphere was in disarray, Cosmic, X-ray, and UV waves would have scorched and challenged cloud formations and the Ozone hole would have had issues, but the Earth survived as did many of the people of that time. From this perspective, it's the calmness of our temperature that seems strange.

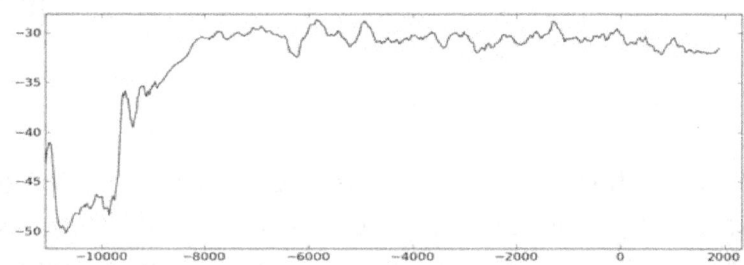

Rather than worrying about building massive solar arrays to generate power for air conditioning, there is another, far more likely, possibility I would worry about. I'm talking about the return to an ice age. As I have shown throughout this book these interglacial time periods only last for 10 to 20 thousand years and this one started 10 thousand years ago. Unfortunately Solar cells may not work well in an Ice Age.

The Next Ice Age

If the current modern global cooling continues, winters in the Northern Hemisphere and summers in the Southern Hemisphere could be colder. The signs seem to be starting already. Greenland data actually shows the temperature is turning colder as the CO2 levels are getting slightly higher as shown below.

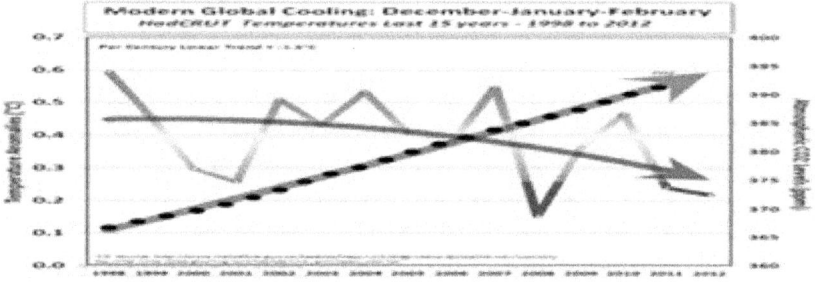

How Much CO_2 is in the Air?

Over the past 50 years, some have speculated that over 1 trillion tons of CO_2 have been thrust into the air by humans burning fossil fuels. As I mentioned before most of these Green Gases would stay in the air for thousands of years so all that trillion is in the air that we are breathing. Even if all this is true, why has there been NO substantial change in temperature cycles? It is now become obvious that the warming trend of the 80s and 90s and even into 2000 has turned around and is now beginning to cool as this massive block of CO_2 just sits there not bothering anyone.

Over the last 10 years the trend has shown a slight drop in temperature. All you hear is the opposite, but whoever cared about the truth.

Thermal Cycles

I showed the next chart earlier concerning both CO_2 and temperature change cyclic nature, but what happens when we just look at after the Pleistocene Extinction?

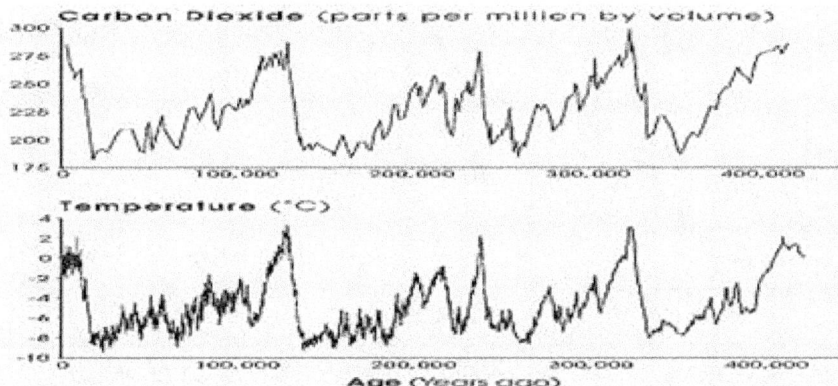

I spread the graph out from about 20 thousand years ago until the present for the Antarctica Ice cores

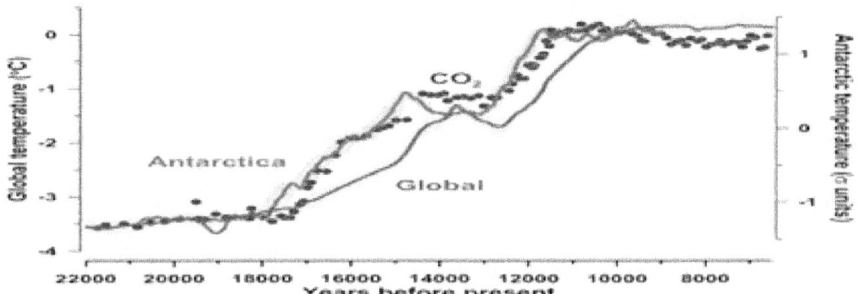

Notice 10 to 12 thousand years ago, the temperature finally stabilized and has slowly been drifting very slightly colder where the Ice core samples were taken near a massive lake toward the middle of the continent. Notice the little dots describing CO_2 levels. Sometimes they are behind the thermal changes and just as many times they are ahead of a

153

change. It is as if Hydrocarbons had almost no effect. When we look at ice core sampling from Greenland we find an even larger anomaly. I superimposed the Antarctica thermal graph on top of the Greenland one. [Antarctic graph is the smaller of the two.] Notice that they are very similar after the Pleistocene Extinction. The temperature actually shows a slight DECREASE in average temperature.

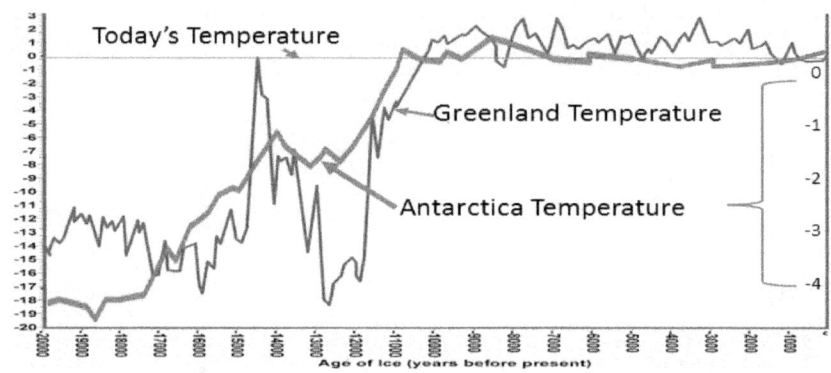

Like I stated before, position of the data is everything. We could get another sample to show a slight increase as well.

Another issue with all of this is that there are massive swings in both directions. The next graph is another Antarctica one showing a very slight increase intemperature over a 400 year period that is less filtered.

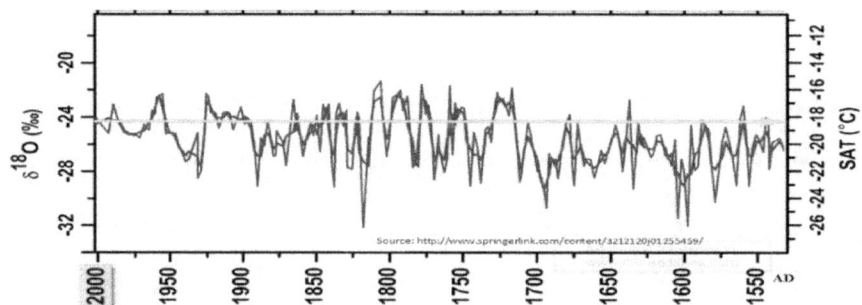

Please notice something about this graph. Nothing seems to change the course of the slope for any period Certainly it shows there were massive temperature drops during World

War II as we began huge manufacturing of warproducts-----
Wait a minute!!! That is the wrong direction as
hydorocarbonated air didn't make the golobal warming
thing. After the war, in the 1950s we see an increase as
massive reduction in manuacturing was noted and so on. If
global warming is not caused from hydrocarbons, what are
we seeing?????? The answer might be on the sun.

*Let's get real! The Earth doesn't seem to be poised to turn
into a pressure cooker. Most likely, what is really
happening is just as bad as global warming. We are
evidently going steadily towards our next Ice Age.*

Many climate experts believe we are overdue for an Ice Age
so they check for indications continuously. The ice core
studies and long term weather research all indicate that the
Earth alternates between Ice Ages and inter-glacials. In
2002, at the South Pole, the penguins were stranded because
of an abnormal ice buildup not because of the blazing heat
the greenhouse model has suggested, but instead the
weather was colder. This is only one of many indicators of
the impending problem. The graph following left is the
deuterium concentration at different depths of ice. There are
two things to notice. The first is that since about 11
thousand years ago, Antarctica has been getting colder. We
have had a base on the continent for decades and have just
had to reposition it as it is now too far from any open sea to
allow for reasonable use. The slope represents about an 8
degree drop in average temperature over that time period.
The only significant longer term thermal events were about
6000 years ago, 10000 years ago, and 11 thousand years
ago [that whole Venus incident]. This brings us to the
second thing to look at. There are thermal spikes that last
for 100 years or less. Although short lived, the thermal

155

change during these times is as much as 3 degrees centigrade. That factor is significant when viewing a later chart.

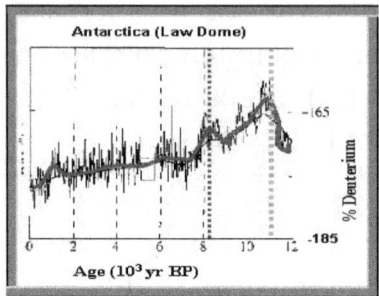
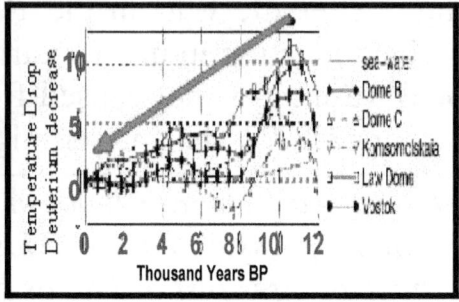

The second chart, right, shows the temperature of Antarctica over 5 different sites. By cross comparing the data, we can be assured that the data has a high degree of accuracy. Antarctica and the entire world **are** getting colder. Instead of the long term ice core graph being shown to us, the short term one below is typically provided to make us think the opposite is happening so we don't use our much needed coal reserves.

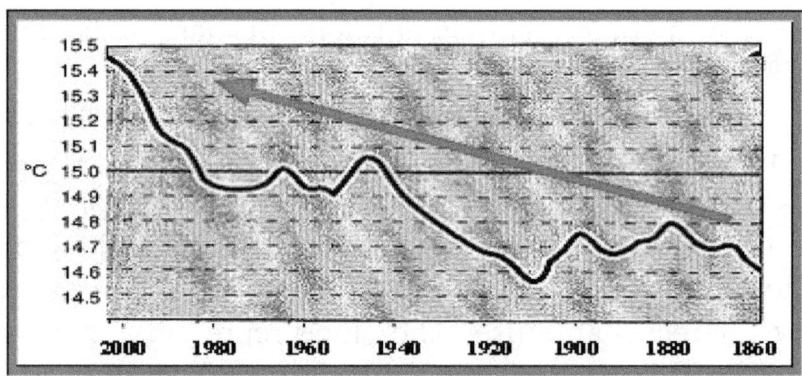

I've been complaining about the halting of all "fluorocarbon sprays" and the testing of "methane producing cattle flatulence" because of the fear of the greenhouse effect, as it is just a way to steer money to "green" industries those

156

jumping up and down have money in. I'll bet you thought that there was "Global Warming" all over, because everyone kept pushing it into your head, but you have been lied to.

This wasn't on purpose, necessarily, but it shows the problem concerning how the science community gets so wrapped up in a theory, that they will do almost anything to show that the theory works.

Here is where there is a lot of conflicting information. Some scientists still hold on to the warming theory and papers are still being written about it every year. If they write enough papers the earth will get warmer, but they will have to write very fast to cause enough friction. Some of the data does suggest a very slight warming cycle, but more of it does not go along with this theory and there is NO data that suggests that anything we do about it has any affect. In all the millions of years of this cyclic behavior, the earth has **never** gone into a significant "Global Warming Fiasco". It's like saying Mars could, all of a sudden, get hot. It won't happen. Our planet has a hard enough time keeping itself warm for us. Remove the atmosphere or disrupt it in any way and the temperatures will plummet. You have seen information in newspapers and articles about the earth warming, but what if we are going the other way? It makes more sense and we need to wake up.

Destroying Trees

Scientists still don't know what even causes Ice Ages much less can they devise ways to slow its progress. We know that the bogus warning for people to halt cutting trees or the earth will get too warm didn't halt tree cutting. The idea of having too much CO_2 because there are not enough trees to consume it is a bad thing, but like the cattle methane, it should not be brought out as a reason for the earth getting warmer. The earth is still doing what it wants to do.

Axis of Rotation

One thing that does absolutely cause the homeostasis of the earth to change is a change in its axis of rotation. If a change like that occurs, plants won't grow as much because of the temperature changes and other environmental changes that disrupt growing cycles. Less coverage of the earth means the earth temperature will drop in temperature significantly rather than increase and large masses of ice would be repositioned from one location to another which would further amplify a delicate situation. This may be one way an Ice Age begins. I showed the hot spot indications of Hawaii and how they line up with major temperature shifts, but around the time of the shifts, there were huge temperature fluctuations as the Earth did not settle quickly in it new position. According to history, the last major shift was only 10 thousand years ago so we are possibly safe from that occurrence for the time being. We are in the region of those violent shifts "after an axis shift". Our Earth is slowly establishing variations on climate associated with the end of the Pleistocene shift. I remember form a movie when Tom Selleck asked a Tibetan Holy Man about what life was. The answer was –

The Oxen is slow but the Earth is patient!

The earth takes a long time to settle. By the way I'm all for halting the extermination of our rain forest, because I like the medicines we get from them, and the oxygen they produce. Fewer trees are not good for any of us. That being said are there signs or axis could shift again?

158

Outer-space Debris and Magnetic Shift

Quite a few researchers are now indicating that our next destruction period will not come from the sky, nor will it come from war or even a simple Ice Age. Our next global event may be another one of those shifts in the rotational axis that caused the Mammoths to freeze about 10 thousand years ago. These scientists have cause to instigate concern as we look at the findings below that are not widely disseminated.

Magnetic Pole Wander

The Earth's magnetic pole has wandered all over the place, and you may think that the wandering was only in ancient times, but we know how dramatic this movement is today because of the work of Paul Serson and Jack Clark, of the Dominion Observatory. The pole wanders daily in a roughly elliptical path around its average position, and may frequently have movements as much as 80 km away from this position when the Earth's magnetic field is disturbed. Accurate observations by Canadian government scientists in 1962, 1973, 1984, and most recently in 1994, showed that a northwesterly motion of the pole is continuing, and that during this century it has moved on average 10 km per year.

The following chart is a running 30-day plot of the readings taken from one of several fluxgate magnetometer sensing sites placed around Canada and Alaska to check for movement of the earth's magnetic field.

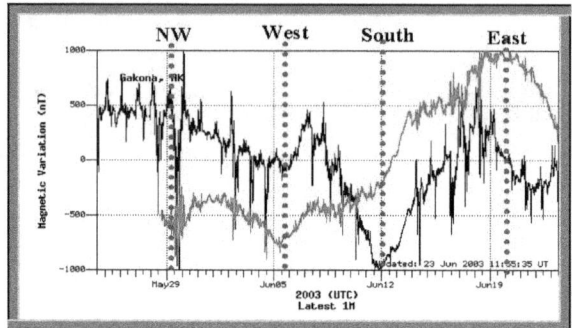

The Upper left component is positive magnetic northward.
The lower left component is positive eastward

Note the wild movement of the Earth's magnetic field as described by this instrument. First the field jumps to the Northwest then moves back eastward, followed by a southward travel and then switching east. Our Polar Satellite data confirms magnetic movement. The graph below shows magnetic wander about the North Pole as captured from space. The general trend towards the northwest can easily be seen.

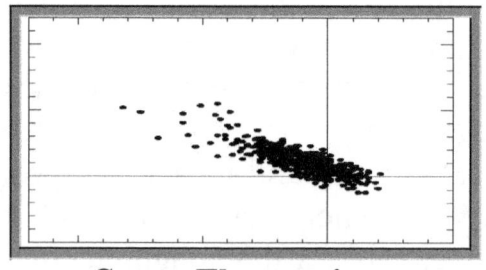

Scary Fluctuations

Since 1994 the average speed of the magnetic drift has **increased** to an average of 15 km per year which makes me sit up a little, but the story gets even more disturbing for

160

those of us who would like to stay in the same climatic position. Satellite data has now been used to compare the strength and direction of the magnetic field in 1980 and again in 2000. According to researchers at "Physics of the Globe Institute" of Paris and the "Danish Space Research Institute", the magnetic field off the southern tip of Africa has already moved dramatically. I know that needs a little explaining. For that we go to Gauthier Hulot, a member of the Danish research team. He and his teammates indicated that molten iron under Africa is now moving in a direction, which will gradually weaken the dominant magnetic field and then reverse it. If the trend continues, the research shows that, we could be seeing the first steps toward a new North Pole. The timing has not been brought out, but saying "any time in the near future" is not good for me. I'm quite used to the hot Florida weather where I live.

Why the Magnetic Field Changes

Researchers have determined that the ultimate cause of the magnetic fluctuations is the Sun. The Sun constantly emits charged particles that, on encountering the Earth's magnetic field, cause electric currents to be produced in the upper atmosphere. These electric currents disturb the magnetic field, resulting in a temporary shift in the pole's position. The distance and speed of these displacements will, of course, depend on the nature of the disturbances in the magnetic field, but they are occurring constantly.

With the earth being generally a sphere and most of the "Iron" component being in a liquid state, the magnetic element of the earth has no particular position it likes or doesn't like. Therefore, it could shift at any time. Whenever the magnetic pole shifts, the rotation axis shift will soon follow as the spinning creates a differential

161

voltage that can only be eliminated by having both elements in the same plane. The shift does not have to be very significant for a destabilizing effect and "Ice Age" to occur so get out your long underwear and buy land in Florida.

Gloom and Doom

I know it all seems like gloom and doom, but all I'm saying is that history repeats itself and the signs indicate that some of the repeating is coming fairly soon. The Earth has gone through many of the following:

Meteor attacks, Axis-of-spin shifts, and Nuclear Wars.

More of each is coming again and putting our head in a hole won't stop it and won't delay it. There may be ways to prepare for some of the disasters, but certainly not if we don't know about them. Take you head out of the sand and make sure that you get the creator's ear, because the signs indicate that time is short.

Meteor Strike

Just like in our past, additional deadly meteors will hit the Earth in the near future. As meteors hit the earth most hit in remote regions and are simply ignored.

Our Last Major Meteorite Strike

No one seems to talk about this event, and it happened in a very remote section of the world, but it could have been very noticeable if it hit anywhere else in the world. The date was **December 9, 1997.** At 5:11 A.M., crews of three trawlers at widely separated sites off south Greenland reported "a blazing fireball that turned night into day." At a distance of over 60 miles away, the flash was compared to that from an atmospheric nuclear explosion. Seismic tremors also emanated from Greenland, so the impact of a large meteorite is almost certain. So far, no one has found the remains of the huge meteorite, but you have to recognize how very desolate and impossible that area is to search. Luckily, our last major meteor hit Greenland and not Disneyland or people would easily accept the event.

The 1997 event was small in comparison to those that will happen in the near future. Before the really big strike there is probably going to be a somewhat smaller one and scientists know about it. The most likely candidate is about 2 Kilometers wide and it is classified as NT7.

Eminent Asteroid Strike

The asteroid 2002 NT7 was just discovered in 2002 and is on an impact course with Earth. It is expected to strike or come extremely close to our planet on <u>1 February 2019</u> and even though it is only 2 kilometers wide, it will be traveling at a rate of 28 thousand miles per hour and contain enough energy to cause continent-wide devastation on Earth, so we should not ignore this terrible threat. To make things worse, if it doesn't hit then, this rock circles the sun every 837 days, so it will have another shot later, but it's not the only chunk of rock that is probably heading for us. To make things look really bad scientists now track a huge quantity of these asteroid masses. The picture below shows those that are close to the earth. Each black dot represents a tracked asteroid. The light circles are the orbits of the major planets and a portion of the orbit of Mars is shown. What I really wanted to show here is that the probability of getting hit is higher than most want to believe. Whether you know it or not, there are many comets and asteroids headed our way and the probability of a significant one hitting is high.

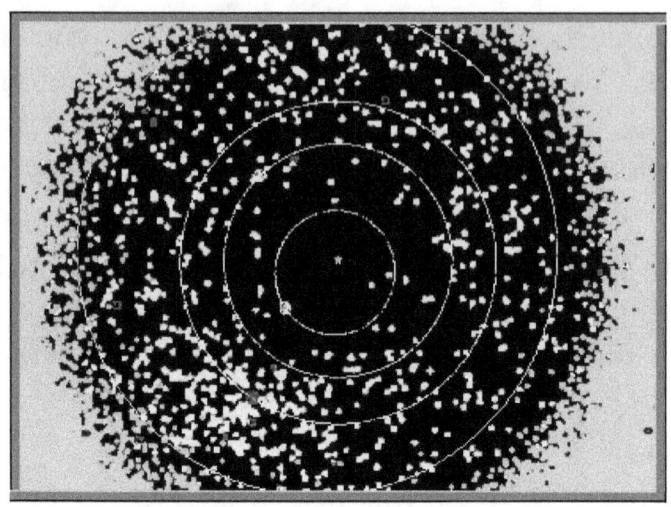

Besides tracked data, we can use two methods that are a little more exotic to find out about upcoming meteoric disasters. The first thing is to know that history repeats itself so something is going to happen in the near future to our weather. Many meteors and even comets have hit in the past and they will hit again. Although tried and true, that method has no timing accuracy or reasoning. The second method is by listening to prophets or seers, but most who claim to foretell the future do not.

Comet Catastrophe

For hundreds of years we have had many people claim to be able to foretell the future. This large group continues to expound on all the tragedies that are upcoming. Luckily most are ignored and as we try to get a better picture of what our future has in store for us, we don't have much choice but to rely heavily on information from some of these Seers provided that a track record of accuracy can be established. That typically means that a seer can't be used to foretell the future until he has been dead and a major portion of his prophecies have come true. We have such a collection and from those few, we may learn that a huge catastrophe is coming soon. By all accounts it will be an attack by a comet or very larger meteor. If history repeats, there will be only one final outcome. The same thing will happen as before. Many will be killed. The meteoric events many times change our weather and much more than CO_2 or even the water vapor we have been discussing.

Dr. Barry Warmkessel and others have provided us with very clear insight into at least a portion of what our future may bring by interpreting 15th century seers, the ancient texts, and one 20th century seer. That future involves comets and meteorites. The information collected and translated includes **what** will happen, **when** it will happen, and **where** it will happen. The following determination was made from prophesies from the Bible, Mother Shipton (1488–1561- England), and Nostradamus (1503–1566- France), and

165

many other stories and traditions. If we want to get a feel for the future, these sources are probably our best hope. Many of the stories concerning an eminent comet strike are integrated within a much more detailed account of our future. In those cases I will try to hold the predictions intact to allow sequential visibility of our world as perceived by the various prophets. When the term comet is used it really is any undesired attack by a rouge planetoid in our Solar System. To start this section, this brief overview comes from the Biblical book of Ezekiel.

Ezekiel's Comet

I'm not going to get into the accuracy of Biblical prophecy, because there are so many predictions that have come true, there can be little doubt about its predictive quality. Many Biblical prophecies tell of an impending comet disaster. This one is in the book of Ezekiel, but others will be provided later.

*__Ezekiel 38-__I will pour down torrents of rain and "GREAT STONES" with burning SULFER on him and on his troops and on many nations. At that time there will be **a great shaking** in the land of Israel.* [The region of Israel being the location of a major disaster and the idea that a great stone causes the damage are both similar to the other predictions that will be provided later.]

When is the Comet Coming?

In addition to the few just presented, predictions of this meteoric disaster occur around the world. Some of the predictions even told us when. The following is a mixture of ancient writings and scientific discussions. We don't know exactly when all the predictions will come about, nor can we be assured that they will, but we may have a better idea about the beginning of the end if we try to cross-compare "prophecies". By these insights, the beginning will be the comet strike. Here is what has been told around the world. Look for the similarities of the impact event and when to expect the catastrophe.

Incan Prophesy- The end of the age of man will be a *Rain of Fire- sometime after 2012.*

Mayan Prophesy- End of the world according to the Maya, or at least, the end of their calendar is ***December 23, 2012.*** *Seems they were off a little.*

Mother Shipton Prophesy- She indicated that the end will be in **2026**. [We will look at her prophecies in more detail]

Nostradamus- *[III-94] For 500 years more one will keep count of him who was the ornament of his time. Then suddenly a great light will be given. "He who is for this century" [Nostradamus] will render them satisfied.* [This strange verse apparently tells us the great light of the comet is coming about 500 years after Nostradamus or in the 21st century.]

Rubinsky- This 20th century seer indicates that an environmental crash will occur **in 2100.**

Zoroastrian Prediction- According to the "Book of Jamaspi", the comet will hit **in 2060**.

Astronomical Events

If we go back to science for a minute, here are the dates of importance.

Meteors- If the comet doesn't hit, there are 2 other major asteroids poised for an impact. The first is named "WN5 and it will pass within 0.00015 AUs [125 thousand miles] in the year **2039.** That is followed by one named "WO107 which passes in 2140 at a distance of 0.0005 AUs [about

167

400 thousand miles]. Both of these predictions are plus or minus 0.01 Aus. [Ouch!]

Photon Belt- According to Astronomers, there is a charged area in the universe, which our solar system periodically enters. Whenever we enter much change occurs. This belt entered our solar system in 2012.

Scientific American- Magnetic field reversed more than 170 times in past 80 million years last reversal, according to "Nature and New Scientist" and "Scientific American" magazines, occurred 13 thousand years ago and the **next shift is expected around 2030AD.**

Biblical Interpretation

Daniel was a great prophet depicted in the Bible. We may glean some details about the coming of this "comet" from his predictions.

Daniel 12:7-*It shall be for a time, times and a half: [2.5 time periods] (If a time period is 1 thousand years it would be 2500 years or between 2000 and 2100.) and when they have made an end breaking in pieces the power of the holy people, all these things shall be finished-* [This verse may indicate that there will be 2,500 years from the time of Daniel until the end of the pre-Tribulation War. We'll talk about the war in a minute, but right now we are using the verse to date the event that sparks the war. This apparently places the comet to come at any time.]

Daniel 12:11- *from the time that the burnt offering shall be taken away [time of Jesus] and the "abomination that makes desolate" [possibly the Moslem rule] is set up, there shall be 1,290 time periods [years] Blessed is he that waiteth and cometh to the 1,335 time periods [years], but go thy way until the end of the days."* [If we assume the time periods are years, it seems to indicate that <u>1290 years after Jesus was executed</u>, a large group against the teachings of Jesus arose and will be influential until the year 2325. If this group is the Moslem nation, then the 1290 AD date indicates the time when the Mongols had taken over much of the Middle and Far Eastern world and had adopted the Moslem religion. This made that "group", the huge power it is today. We will discuss the Moslems in more detail later. If this is the meaning, then a major war will be ended by about 2325AD.]

There is no doubt that if a significant meteor or comet hits, our weather will change, the magnetic field could fluctuate enough to shift our axis again, or the disruption of our clouds and atmosphere could force a new Ice Age.

Mother Shipton's Revelation

Not only did some of these seers help us understand about how our Earth would react in the future, but also how we will adapt to the environment. First, let's look at this Mother Shipton character. Her real name was Ursula Sontheil and she is a reasonable prophet to use because of her accuracy. Her prophecies have had an extremely good record for coming true just like those of the more famous Nostradamus. Here is a short list of extremely detailed and accurately predicted historical elements that she wrote about. People thought that she was a witch and eventually killed her, but before they did, she had unveiled the future. [By the way she accurately predicted the method and timing of her own lynching. I think it is better not knowing what the future will bring, sometimes. She predicted the following:

- Automobiles,
- The rise of the Church of England,
- Radios, telephones, telegraphs, hydroelectric power,
- Manufacture of mountain tunnels,
- Submarines, airplanes, & iron ships,
- The California gold rush.
- World War I

- US Civil War and the French Revolution.
- Airborne military and their use,
- British and French alliance during the World War,
- The Allies and Communist bloc, and the cold war,
- The France to England underwater tunnel,
- Women would commonly wear pants and have short hair, [an unthinkable thing at the time]
- Commercial air travel,
- Assemblies would be put together with huge machines,
- The printing press and how it would change writing forever.

Shipton's Comet

With all of that as verification, I think we can use her other predictions of still more distant future events. The following set of predictions is supposed to occur during this current century starting in 1926. I don't know why she started with the odd year, but it means that she believed that all of the following will occur within the next 20 to 25 years.

For those who live the century through *- in fear and trembling this shall do. "Flee to the mountains and the dens-to bog and forest and wild fens. For storms will rage and oceans roar when Gabriel stands on sea and shore, and as he blows his wondrous horn old worlds die and new be born.* **[This catastrophe will happen just before the year 2026]**

God's messenger from the heavens *(comet) arrives and a great sound is heard as it passes through Earth's atmosphere and impacts Earth. It causes wild storms and raging seas.*

A fiery dragon *(a comet as marked by it "dragon tail") will cross the sky six times before the earth shall die. Mankind*

171

will tremble and frightened be for the six heralds in this prophecy. [**It could mean six major meteor strikes will occur; spawned by the comet strike.**]

For seven days and seven nights *man will watch this awesome sight. The tides will rise beyond their ken. To bite away the shores and then the mountains will begin to roar and earthquakes split the plain to shore.* [**The strikes happen over a seven-day period or on the 7th day.**]

And flooding waters *rushing in will flood the lands with such a din that mankind cowers in muddy fen and snarls about his fellow men. He bares his teeth and fights and kills and secrets food in secret hill and ugly in his fear, he lies to kill marauders, thieves and spies.* [**People begin killing each other over food.**]

The world upside down *shall be, and gold found at the root of a tree.* [**This is saying that the Earth axis will shift either before or after the comet strike. It is not known when. Certainly this will greatly affect our atmosphere like it has in the past.**]

Yet greater sign there be to see as man *nears latter century. Three sleeping mountains gather breath and spew out mud, and ice and earth and earthquakes swallow town and town.* [**Her centuries ended on the 26th year after a normal turn of the century, so she was talking about the time between 2001 and 2026. Earthquake and Volcanic action becomes significant. Both could be caused by the comet or by the upcoming Earth axis shift.**]

Pre-Tribulation Dark Age

Mother Shipton put her predictions in a simple package. Her predictions follow many of the others completely, but they add their own dimension. Possibly the cause of the

172

great Tribulation Plague era is a mistake made by scientists and possibly there is involvement from outside the earth.

Famine

Man flees in terror from the floods and kills, and rapes and lies in blood and spilling blood by mankind's hand will stain and bitter many lands. And when the dragon's tail [comet] is gone man forgets and smiles and carries on. To apply himself too late, too late for mankind has earned deserved fate. **[After the comet, blood is spilled by war.]**

Moslem War

War will follow with the work in the land of the Pagan and **Turk** [Indicates that the Turks will eventually follow the Moslem nations in this pre-Tribulation War. This will become clearer in other writings.]

Moslems are Beaten

The **lily [USA?]** *shall be moved against the seed of the* **lion [sign of Persia]**, *and shall stand on one side of the country with a number of ships.* [Lily to the rescue. The Moslem lion is beaten back by a nation with a strong Navy, possibly USA.]

The Beast Controls the World

With a number [possibly 666] *shall he* ***[This "son of Man character"]*** *pass many waters and shall come to the land of the lion* **[Persia]**, *looking for help from the* **beast** *of his country* [Some group goes a great distance across the water, probably the USA, to fight the Moslem Lion. Unknown to them, they will be aided by the beast; what Nostradamus called the Prince of Hell.]

The lily F.K. shall lose his crown, *and therewith be crowned the Son of Man K.W. and the fourth year shall be preferred.* [Evidently the United States "Lily" and the Son

173

of Man "Prince of Hell" rule the world together for a time, but the old Prince takes full control after 4 years. Shipton even told us something about them with the peculiar initials. If someone becomes president with the initials FK I'm going to start worrying.]

Tribulation Period

These mighty tyrants *will fail to do. They fail to split the world in two. But from their acts a danger bred, an age - leaving many dead and physics find no remedy for this is worse than leprosy.* [Here is where the real doom and gloom starts. The two leaders of the world have a bitter struggle and this seems to indicate that scientists cause the plagues of the tribulation while trying to find ways to gain control. She indicates there is no cure for what is unleashed.]

Who survives this terrible din *and then begin the human race again. But not on land already there, but on ocean beds, stark, dry and bare.* [Terrible plagues erupt and no rain causes much of the rivers and sea dry up during this time, just like the Biblical prophecy indicates.]

Not every soul on earth will die*, as the dragon's tail* ***[Comet]*** *goes sweeping by, not every land on earth will sink, but these will wallow in stench and stink of rotting bodies of beast and man, of vegetation crisped on land.* [By the end of the Tribulation Era, the world is a mess.]

World Peace

And before the race *is built anew, a* ***silver serpent*** *comes to view and* ***spews out men of like unknown*** *to mingle with the earth now grown cold from its heat* [Seems to be indicating that the Earth climate will change and become cold. Somehow there is a way to restore the Earth environment.

And these men can enlighten the minds of future man to intermingle and show them how to live and love and thus endow the children with the second sight; a natural thing so that they might grow graceful, humble and when they do the golden age will start anew. [These visitors also teach the remaining people. This new era is predicted in the Biblical version as the 1000 years of peace.]

And there shall be a universal peace over the whole world, and there shall be plenty of fruits; and then he shall go to the land of the Cross. [As in the Biblical prediction, during the time of peace, God {land of the cross} will reign.]

I know some of the interpretations seem a little weak when only looking at the Mother Shipton version only, but just wait until you read the versions provided by John in "Revelation" and the Nostradamus Quatrains. Before we get into the Biblical revelation let's look at the sequence of events again. [See below] Compare this sequence to the others as we look for cross comparative similarities to assure truth.

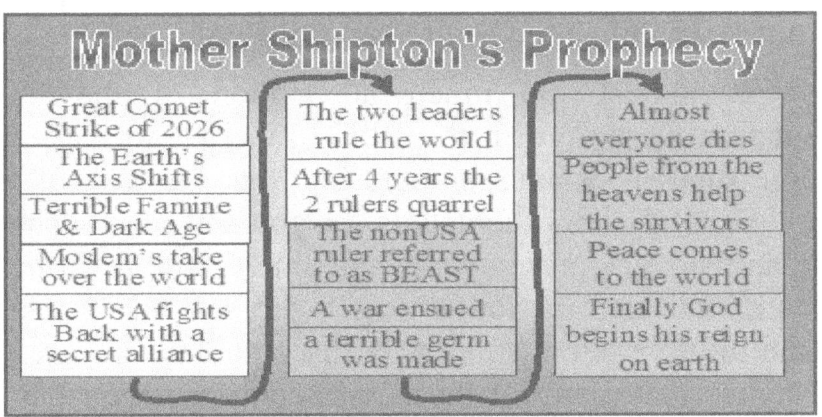

John's Revelation

Mother Shipton's prediction is nothing compared with that of John in the Biblical book of Revelation, but there is a substantial amount of similarity and that is what we really need to focus on. In order to establish this version of "Revelation" I took the information sequentially. I know that is generally unheard of in interpretation of this masterful document, but it made the most sense to John and it makes the most sense when comparing the document to other information. While many do not bring it up, the similarities begin with a bang. Just like in the Shipton predictions, the events of our immediate future begin with a comet or meteor strike according to John's vision. The meteor strike will signal the chain of events that lead to "the" final battle according to this, best known, prophet of our future. It will be the beginning of the end.

There is almost too much description of the future Tribulation Dark Age in the Bible. In order to make this section smaller, I am paraphrasing the verses, but you should certainly read the details for yourself as you should with any details presented by anyone. John describes a vision of this time period as angels, horses, and trumpets, but the information is still easily discernable. I'm sure you will notice that many of the items seem to repeat, but the sequences are slightly different for each era so we can be pretty sure that the periods being described are different for each section. Before we look at the actual texts, I have made out a chart showing the timeline of the Tribulation Dark Ages according to the Book of Revelation that can be used as sort of a guide to his prediction. [See next graphic] You will recognize a similarity between the Shipton prophetic timeline and John's right away, however, John's story is separated into three distinct sections; each of the 6 events identified in each section is signaled by a symbolic article such as a seal, trumpet, or vial. Each section also becomes progressively shorter in time than the preceding one. In addition to getting shorter, each "era" is progressively worse for humans. After this review, we will look at other stories to show consistency and comparison, but this is certainly the most complete description. Between the second and third eras, there is a series of wars that are described in fairly fine detail and the third era is very detailed in that it only covers a period of 3 ½ years.

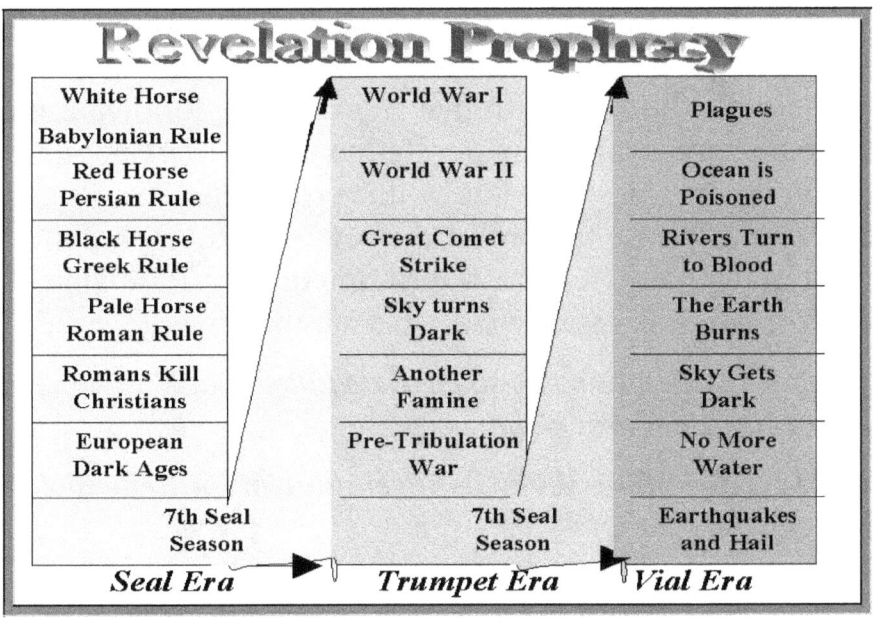

Revelation Prophecy

Seal Era	Trumpet Era	Vial Era
White Horse / Babylonian Rule	World War I	Plagues
Red Horse / Persian Rule	World War II	Ocean is Poisoned
Black Horse / Greek Rule	Great Comet Strike	Rivers Turn to Blood
Pale Horse / Roman Rule	Sky turns Dark	The Earth Burns
Romans Kill Christians	Another Famine	Sky Gets Dark
European Dark Ages	Pre-Tribulation War	No More Water
7th Seal Season	7th Seal Season	Earthquakes and Hail

The "Seal" Era

The first era is marked with seven sequential events, each represented by a seal. From this first group, we should get a very strong wake up call. As many of you already know and as noted in the chart the first four "seasons" of the "Seal Era" are marked by a different color horse to amplify their meaning. The others are not so amplified, so there is some particular emphasis that is desired concerning these events. I know that people continue to talk about the 4 Horses of the Apocalypse, but what you should initially think is very strange is the fact that there are not 5 or six horses representing the other seasons. I think most people have gotten it all wrong.

Why Are There Horses?

There is a very good reason why the first 4 seals were depicted as horses and it has nothing to do with the future as you have been told. The reason the first 4 seals are turned into horses is that the seals had already been broken by the time of John, so they aren't part of our apocalypse now. They are just horses. With that in mind, the following may be the most reasonable interpretation of the Revelation. We will compare these events with prophecies from other sources to increase the probability of correctness, but this is our starting point for the time immediately following the comet strike. Let's start with horse number one.

Babylon Rule [600-500BC]-*Revelation 6:2: The first season [White Horse] -the world begins to unite.*

Persian Rule [500-300BC]-*Revelation 6:3-4: The second season [Red Horse] - war breaks out everywhere.*

Greek Rule [300-50BC]-*Revelation 6:5-6*- *The* **third season [Black Horse]** - *an intolerable famine occurs during the wars.*

Roman Rule [-50 to 500AD]-*Revelation 6:7-8 The* **fourth season [Pale Horse]** - **Power was given to them over** ¼ *of all people will die.*

Mislead European Age [500-1000AD]-Revelation 6:9-15 -The fifth season - Many God worshipers will be slain during this time.

The European Dark Age [1000-1500AD]-Revelation 6:16-The **sixth season**- Stars will fall; Great earthquakes will occur. Kings will hide themselves in fear.

The Age of Enlightenment [1500-1900]-Revelation 7:1- The **seventh seal** indicates that a new era brings similar events identified as trumpets.

The "Trumpet" Era

The second era is also marked with seven sequential events that occur immediately after the first era. Each of the seasons is represented by a trumpet and are all played out within the last season of the "Seal Era". The first 6 seals took about 3000 years to complete. The next 6 will take considerable less time; perhaps only about 500 years or until the 21st century. The significant event for us today is the events after the third trumpet which apparently is a major comet strike.

Revelation 8:1 The **seventh season** *divided into 7 more seasons represented by trumpets. Therefore we know that this first season is 6 times longer than the remaining time.*
World War I-*Revelation 8:7-The* **first trumpet Season-***A hail of fire and blood. One third of the world will be burnt*

up. [**Sounds like World War I where <u>1/3 of the world was fighting</u>.**]

World War II-Revelation 8:8-9-The **second trumpet season-** a third of the living creatures in the sea will die, and 1/3 of the ships will be destroyed. [**Sounds like World War II to me with its submarine warfare. As I mentioned before, hundreds of thousands of metric tons of ships were sent to the bottom of the Ocean.**]

Time of the Comet [2020?]

<u>Revelation 8:10-11</u>-The *third trumpet season-* *a huge comet falls and contaminates 1/3 of all the drinking water.*

<u>Revelation 8:12</u>-The *fourth trumpet season-* *the sky will be darkened for a time*

<u>Revelation 9:1-11</u>-The *fifth trumpet season-* *another famine lasting five months.*

I put these together, because this time of famine and darkness might not last very long. The next trumpet is right behind this one.

I know this sounds identical to Mother Shipton and we will find that Nostradamus tells us the same thing. A comet hits, something happens to our environment, the famine as we have a time to regulate the Earth.

The Pre-Tribulation War

Again we find war during time of hardship and like the others; the War will shift to a Moslem invasion and finally a time for them to be beaten back.

<u>Revelation 9:13-19</u>-The *sixth trumpet season-* *third of mankind will be killed.*

<u>Revelation 11:2-5</u> *The* *holy city shall they tread under foot forty and two months. And I will give power unto my two*

witnesses, and they shall prophesy a thousand two hundred and threescore days. And if any man will hurt them, fire proceedeth out of their mouth, and devoureth their enemies: **[This seems to indicate that 2 leaders will push back the Moslems after another 3 ½ years of Moslem rule. If you remember Mother Shipton identified these two guys as F.K. and K.W. I don't care who they are if they really push back these aggressors quickly.]**

Moslems Take Control

Revelation 12:3- *and behold a great red dragon, having seven heads and ten horns, and seven crowns upon his heads.* **[Note the difference between this first depiction and the one presented in chapter 13, below. This one has only 7 crowns while chapter thirteen indicates that that 3 more crowns were picked up, but both this verse and the following one are depicting the same thing. It is a group of ten non-Christian nations going against the Christian nations of the world. From other texts we can clearly see this is talking about the 10 Moslem nations.]**

Revelation 13:1 *And I stood upon the sand of the sea, and saw a beast rise up **out of the sea**, having seven heads and ten horns, and upon his horns ten crowns, and upon his heads the name of blasphemy.* **[We will see that ten non-Christian nations will attack the Christian nations by sea, just as this verse predicts and we will find confirmation in the writings of Nostradamus and others. The ten horns are of extreme importance as are the 7 crowns and 10 crowns.]**

Revelation 13:7 And it was given unto him [the Moslem nations] to make war with the saints, and to overcome them: and power was given him over all kindreds, and tongues, and nations. **[The reason that the Moslem** nations have been indicated as the aggressor will be much more evident in the section from Nostradamus and if you recall from the Mother Shipton section the aggressors were termed as those coming from Persia which would be the same group.

Nostradamus' Revelation

His name was actually Michel de Nostradamus. Before we get into this guy's revelation, let me tell you something you might not know about him. Nostradamus means "Our Mother" which certainly would have referenced Mary, the mother of Jesus. This is a very odd name until you remember that, in the late 15th century; Jews were being hunted down by the hundreds and killed for no other reason than they were not Christian. It would be reasonable that if you had a Jewish family name, you had better change it quickly or you could have been dog food. Michel's father quickly found a French girl to marry from a respected family, but King Louis VII still came up with an ANCESTRY TAX. Nostradamus' father had to pay for changing his nasty name into the nice French name he had adopted. I'm sure there was resentment in the Nostradamus household, but the hiding and taxing was better than getting killed. Anyway, old Michel de Nostradama was no slouch as he learned the ways of Astrology from his grandfather on his respectable "French" side. He complied hundreds of verses that appear to be very accurate predictions of his future. Many of these verses have, Apparently, come true as indicated below.

- He predicted Hitler's reign and called him by the name of Hister.

- He called out events associated with and correctly named Franco
- He called out Mussolini's reign and alliance with Germany.
- **He indicated that twin towers would be attacked from the sky that occurred September 11, 2001 in New York.**
- He Named and dated the great fire of London.
- He had many predictions concerning Napoleon which came true and many, many, many more.

September 11 2001 Attack

Here is what he had to say about the twin towers attack:

*A New City on the 45th latitude [**Guess where New York is!**] is attacked with fire from the sky as the North is put to the test. The earth shaking fire from the "World's Center" burns around the New City, two rocks make war on each other. The eagle-like attacker [**Jet**] of the New City is at first uncertain, then magnanimous victory, with damage to Cremona and Mantua* [**These are the names of two giants like the twin tower giants**].

Comet Attack

I mentioned that Nostradamus provides us with details and the comet event is a good place to start. Nostradamus tells us of a similar comet catastrophe to that of the preceding stories, but in this case he may have even told us where the meteors hit. His prophecy is so descriptive; we can build a map and track the invader. He tells us when it will hit, the size of the rock, and where the pieces will land.

*Quatrain II: 46- After great human misery, a greater one approaches, The great motor of the **century renews**: It will rain blood, milk, famine, iron and pestilence. In the sky will* *be seen a fire, dragging long sparks.* - *[A large meteor strikes a short time after the new century. They will introduce famine and disease.]*

Quatrain I: 69-*The great mountain, 4247 feet in circumference, After peace, warmth, famine, And flood: The impact will spread far, drowning with great oscillations, even ancient objects and their great foundations.* [**The diameter of the meteor is 400 meters. It causes great tsunamis. Even very ancient objects are destroyed.**]

Quatrain VIII: 16-*At the place where JASON had built his ship, There will be such great and so sudden a flood, That one will have nowhere on earth that was not attacked.* **[Floods are felt around the world, but possibly Greece is hit first.]**

Quatrain VI: 6- *There will appear towards the seven stars. Not far from Cancer, the bearded star:* **Susa, Siena, Boeotia, Eretria, Great Rome** *will die, the night disappeared.* [**Possibly indicates that the destroying comet will first appear near the constellation cancer. Again Mediterranean cities are mentioned on the destruction path.**]

Quatrain III: 10-*Greater calamity of blood and famine. Seven times it will advance toward the marine shore:* **Monaco**, *from hunger, place captured, captivity, the great leader is crushed in a metal cage.* [**The comet splits into 7 pieces before hitting along the shoreline near Monaco.**]

Quatrain I: 46-*Very near to* **Auch, Lectoure and Mirande**, *Great fire of the Sky in three nights will fall; which will cause a stupendous event to happen. A short while after the Earth will tremble.* [**This verse indicates that one meteor**

strikes near the three French towns mentioned. The strike initiates huge earthquakes.]

Quatrain V: 98-*At latitude* **of 48 degrees** *climatic. At the end of Cancer so great is the drought: Fish in the sea, river and lake boiled hectic, Bearn, Bigorre the sky in distress from fire,* **[In western France, a great drought is felt from the fire in the sky.]**

Comet Strike Plot

From the above predictions and the interpretive work of researcher Barry Warmkessel, we can make out the destruction path. The picture below shows the intersection of the prophesized hits- Susa - Italy/France Border, Siena - Tuscany Italy, Eretria – Greece, Monaco, Boeotia – Greece, Mount of Olives- Jerusalem, Rome-Italy, Auch, Lectoure, Mirande, Bearn, Bigorre-France, and Sardinia.

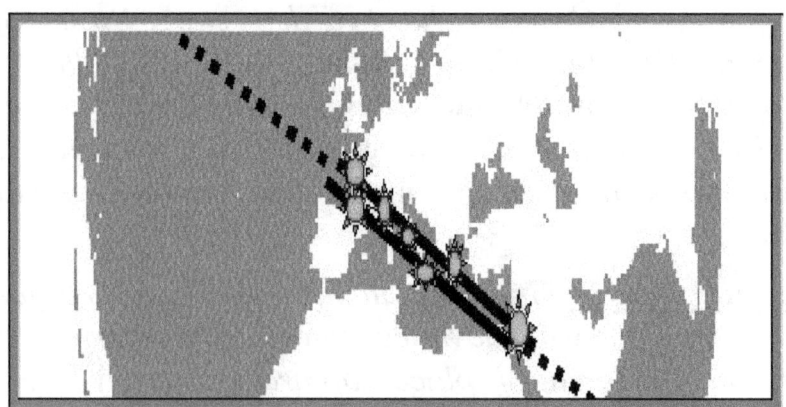

Pre-Tribulation Dark Age

Like John and Mother Shipton, Nostradamus saw a huge shift in our environment bringing a great European famine and plague. In its weakened state, Europe was ripe to be overrun and a Moslem Nation alliance soon takes advantage. This goes along the predictions in Revelation and other Biblical texts completely, but that is of little comfort. He even times the event as being in the 21st century.

[VIII-16]At the place where Jason had his ship built [Greece], there will be a flood so great and so sudden that one will have no place to fall upon. The waves of Olympian "Fesula" [Roman Island] **[The Comet strike will bring a tidal wave and that will be only the beginning.]**

[II-96]A burning torch will be seen in the sky at night near the end and the beginning of the Rhone [southern France]; Famine, then sword, then relief provided too late. Persia turns to invade Macedonia **[After the comet, there will be a terrible famine in Europe. The famine weakens Europe enough to allow the Moslem nations to strike Greece.]**

[IX-91]The horrible plague will fall upon Perinthus [on the Turkey/Greece coast] and Nicopolis [Greece], the peninsula and Macedonia. It will devastate Thessaly and Amphipolis [both in Greece]. It is an unknown evil and from Anthony a refusal. **[A plague also weakens the already weaken European nation. Possibly Moslems try to make a surrender deal, which is refused. That paves the way for war.]**

During the third age, fire from the sky spreads from west to south toward the east, sterilizing the soil and leaving behind a country littered with glowing "carbuncles" and laid waste by famine. **[More descriptions of the aftermath of the comet]**

[VI-5]A very great famine occurs after a pestiferous wave [plague], through a long rain of the Arctic pole. Samarobryn [possibly Nephilim], one hundred leagues above the hemisphere; they will live without law, exempt

187

from politics. [**While the plagues and famine happen as a result of the comet strike, some beings are outside the earth contemplating the attack on heaven identified in Revelation. They live without the laws of God or man, so they are doing something not desired by either.**]

Moslem War

<u>**[II-45] The man-woman cries**</u> *in the high heaven, where blood is sprayed. They avail too late and a mighty race dies. Sooner or later the hoped for aid comes.* [**War in the high heaven essentially kills the once mighty race. One can assume it to be some remaining Nephilim although it could be that group of humanoids I mentioned that might be living on the moon.**]

<u>*[I-40]The false trumpet*</u> *concealing the madness will bear Byzantium [Turkey] a change of laws. From Egypt one will go forth who wants withdrawal of the edicts altering money and standards.* [**Just like the Revelation prediction the last trump signals in a great pre-tribulation war. Turkey and Egypt are reluctant, but go to the non-Christian faction**]

<u>*[V-25]The Arab prince*</u> *will come during the time of "Mars, Sun, and Venus in Leo"* [**next alignment will be in 2020/1**]. *The rule of the church will be succumbed by sea.* [**This indicates that the initial attacks are probably going to come from the Mediterranean Sea.**]

<u>*[VIII-96]The barren Synagogue*</u> *[Israel] will be without fruit. Taken over by the infidels. The daughter of the persecuted Babylon will be miserable and sad. Her wings will be clipped.* [**After the Moslem Nations Unite, Israel will be taken. {What a shock!}**]

[IV-58]To swallow the burning sun in the throat, the Etruscan land [Italy] is washed with human blood. The chief uses a pail of water to lead his son away. A captive lady conducted into the Turkish land. **[The Turks hit Italy while they are still riling from the famine.]**

[II-30]One whom the infernal gods of Hannibal [Libya] will cause to be reborn, terror of mankind. In the past, no more horror nor worse days are than will come to the Romans [Italy] through Babel [Iraq]. **[The Iraqi will align with Libya, to take over Italy.]**

[III-60]Throughout all Asia [Asia minor] there will be a great buildup of troops even in Mysia, Lycia, and Pamphilia [All in Turkey]. Blood will be shed because of the absolution of an evil young king filled with felony. **[Again we see a reluctance of Turkey to go with the non-Christians.]**

[III-64]The chief of Persia [Iran] will engage great Olchades [Spain]. The three-pronged fleet against the Mahometan [Mohamed] people comes from "Parthia [ancient Iran] and Media" [ancient Iran]. The Cyclades [Greek Islands] are pillaged. Long rest at the great Ionian port. [Greece] **[Spain will fall along with Greece.]**

[V-55]In the country of Arabia Felix [Saudi Arabian ruler?], there will be born one powerful in the law of Mahomet [Mohamed]. Buy sea, he will vex Spain, conquer Genada [Spain], and further, by sea, the Ligurian people [Italy]. **[Possibly this guy is the "lily F.K." that Mother Shipton said would shall "lose his crown", and therewith be crowned the Son of Man K.W.]**

[VI-80]Through Fez [Morocco], the realm will reach those of Europe. Their city ablaze and the blade will cut. The

great one of Asia will come by land and sea with a great troop. Blue and Perses [dark blue] will come and will pursue the cross to the death. [**Here come the blue turban thing we always here. If Morocco is taken, Europe better watch out.**]

[VIII-6]Naples [Italy], Palermo [Spain], and all of Sicily will be uninhabited because of a Barbarian [non-Christian]. Corsica, Salerno, and the Island of Sardinia will find famine, plague, and war. The end of evil is remote. [**The devastation is bad and the aftermath of famine and plague lasts a long time. Italy, Spain, and Sicily seem to be the hardest hit.**]

[I-73]Because of negligence, France is assailed in 5 places. Tunis and Algiers [North African coast] are stirred up by the Persians [Iran]. Leon, Seville, and Barcelona will fail leaving no fleet for the Venetians [Italy].

The graphic below shows how the invasion, apparently, proceeds.

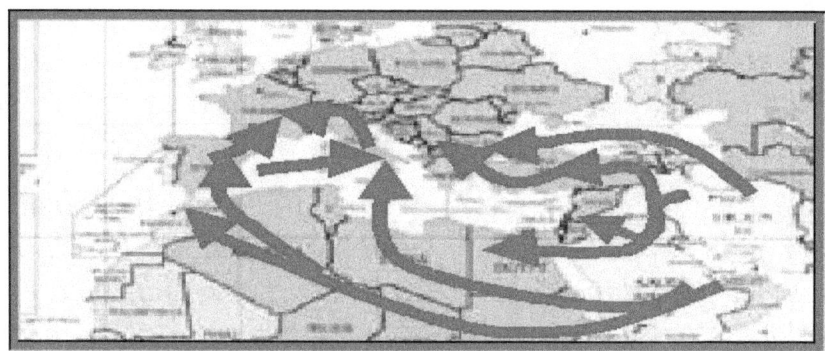

The Moslems are Pushed Back

These words speak for themselves. The Christian forces, initially from Russia and German, push the Moslem nations back, but it isn't easy and there seems to be reference to a nuclear attack.]

[X-75]After time he will be pushed out of Europe. He will be sent back to Asia. As one of the League from the great Hermes [Greek messenger of the gods], he will grow over all the kings of the East. **[Possibly indicating that a mighty leader will come from Greece.]**

[V-27]Through the fire and arms not far from the Black Sea, he will come from Persia [Iran] to occupy Trebizond [Turkey/Russia border]. Pharos [Egypt] and Mytilene [Greece] will tremble. The Sun will be joyful. The Adriatic Sea will be covered with Arab blood. **[This is pretty clear, the Iranian Army tries to go into Russia and they don't make it. If that isn't clear enough, the next one says it again.]**

On the fields of Media [Iran], Arabia, and Armenia [Russia/Turkey border], two great armies will assemble three times. The host near the banks of the Araxes [border of Iran and Armenia]; they will fall in the land of the great Suleiman [Turkey]. **[Russians push back the Moslems.]**

Towards Persia very nearly a million men, the true serpent will invade Byzantium and Egypt. **[This portion of a verse talks about the true serpent marching towards Persia, Turkey, and Egypt. This apparently predicts Moslems being pushed back by an even worse demon. As we found out from Revelation, this new serpent uses the number 666.]**

[V-16] Flesh will be turned into ashes through death, at the island of Pharos [Egypt]. They will be disrupted by the Crusaders [Christians] when a harsh specter will appear a Rhodes [Greece]. **[This harsh specter appears to be Nuclear.]**

[V-74] A Germanic heart will be born of Trojan blood [Turkish?]. He will rise to very high power. He will drive out the foreign Arabic people and return the Church to its pristine preeminence. **[This time it seems that the Germans are on the winning side of the battle lines and attack Italy. Also note that the German leader, apparently, has descendants from Turkey.]**

A final battle just to the northeast of Salon leads to the collapse of Mesopotamian [Iraqi] power in France. **[Iraqi soldiers are pushed out of France.]**

After tarrying, they will venture into Epirus [northwest Greece]. The great relief will come towards Antioch [Turkey]. The black crimpled-haired king [Moslem] will run quickly towards the empire. The "Bronzebeard" will roast him on a spit. **[After the "great relief" [possibly USA?] comes, the Moslems are quickly pushed out of Europe. The old "Bronzebeard" guy is not known to this author.]**

[IV-85]The great city of Tarsus [Turkey] will be destroyed by the Gauls [Germans]. All the turbans [Moslems] will be captives. Help by sea will come from the great one of Portugal during the first summer, St. Urban's Day. **[I don't know when St. Urban's Day is, but I suppose it will become a new Holiday of victory.]**

The graphic following shows how the Moslems, apparently, are pushed out of Europe.

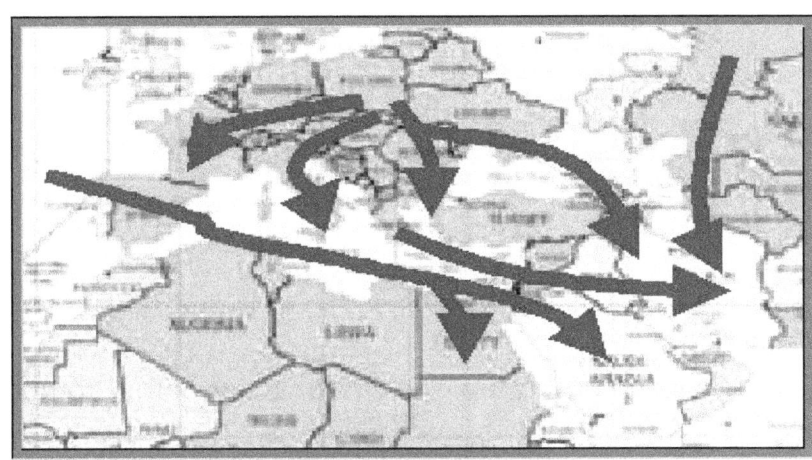

Mark of the Beast

Nostradamus tries to clear the confusion between the Moslem horde leader and a new nemesis that takes control of the world after the Moslems are pushed back. He, apparently, called the Moslem leader "antichrist" and he is completely separated from the 666 guy that Nostradamus calls the "Prince of Hell". He even gave us a time period for the Prince of Hell to be in control. His prediction is short and to the point. After the Antichrist, Nostradamus indicated that the Prince of Hell will reign over the world spreading terror and universal destruction for twenty-five years. One place he calls him the King of Terror, but he seems to be the same entity.

From Heaven will come a great King of Terror to bring back to life the Great King of Angolmois [possibly an anagram of Mongol] before and after Mars to reign by good luck. Evidently the end of the 25 years is the Tribulation period discussed previously followed by peace.

Esdras Revelation

The 4th book of Ezra in the King James Version of the Bible is also known as the 2nd book of Esdras. It provides us with a wealth of information concerning our future and, like the book of "Revelation" and the other prophecies already examined, the details are probably not what we would like to hear. Again we find wars, more wars, famine, meteors, and a single governor for the world. The timeline is slightly different than that found in Revelation, but the elemental parts are similar.

Esdras Credentials

In addition to these future predictions, Esdras also predicted correctly the coming of the Roman Empire, its twelve leaders before Augustus and the terrible times experienced by the Jewish community during the future for him, but the past for us today. An interesting element of this future prediction is that it either is retold 4 times or there are 3 "cycles" much like the 3 cycles described in "Revelation". A fourth cycle is also predicted to be similar to "Revelation" ending with a final War. If we place the predictions in order, the events are as shown on the following graphic. They don't look fun, but I never said this was a comfortable history.

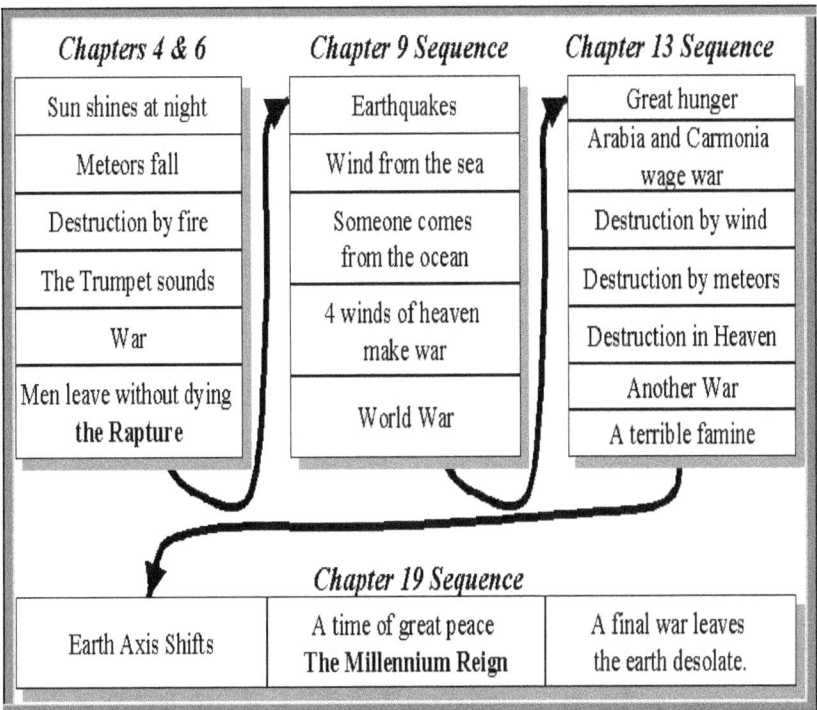

Season One

[4:1]__The days are coming__ when those who dwell on earth shall be seized with great terror.

[4:3] __And the land which__ you now see ruling shall be waste and untrodden, and desolate.

[4:4] __The sun shall suddenly__ shine forth at night, and the moon during the day. [Possibly, the Earth shifts its rotational axis before the next major comet attack.]

Comet Strike

[4:5] __Blood shall drip__ from wood, and __the stars shall fall__. *[Meteors fall]*

[4:6-7] __And one shall reign__ whom those who dwell on earth do not expect, and the sea of Sodom shall cast up fish; **[Interesting prediction, but I'm not sure what it means, but something bad happens to the seas.]**

195

*[4:8] **There shall be chaos** also in many places, and fire shall often break out,* [**Destruction by fires**]

*[6:23] **and the trumpet** shall sound aloud,* [**This is the "Last trumpet sounding" as we saw in "Revelation".**]

Pre-Tribulation War

[13:3] The figure of a man came up out of the heart of the sea. That man flew with the clouds of heaven; and wherever he turned his face to look, everything under his gaze trembled. [**Seems to be saying that someone will come out of the ocean to take control during this time. Before he takes over he, apparently spends sometime in the heavens.**]

*[13:5]**After this, an innumerable multitude** of men were gathered together from the **four winds of heaven** to make war against the man who came up out of the sea.* [**This strongly suggests that those who would war against this leader came from the "4 winds of heaven"**]

*[13:8] **All who had gathered** together against him, to wage war with him, were much afraid, yet dared to fight.*

*[13:17-31] **those who are not left** shall plan to make war against one another kingdom against kingdom.*

*[15:13] **Let the farmers** that till the ground mourn, because their seed shall fail and their trees shall be ruined by blight and hail and by a terrible tempest.*

*[15:15] **Nation shall rise** up to fight against nation, with swords in their hands.*

*[15:19] **A man shall have no pity** upon his neighbors, but shall make an assault upon their houses with the sword, and plunder their goods, because of hunger for bread and because of great tribulation.*

Moslems Take Europe and are Driven Back

*[15:29-33] The nations of the dragons of Arabia shall come out with many chariots, and from the day that they set out, their hissing shall spread over the earth. The **Carmonians**, raging in wrath, shall go forth the dragons, remembering their origin, shall become still stronger; and if they combine in great power and turn to pursue them. Then these [the dragons of Arabia] shall be disorganized and silenced by their power, and shall turn and flee.* **[It appears that the wars during this future time will be initiated by the Middle East and finally taken over by the Carmonians who have joined with many other nations. I chose to assume Carmonia means the USA, but that is just me.]**

*[15:34] Behold clouds from the east and from the north unto the south **[South of Israel or the Arabs]**, and they are very horrible to look upon, full of wrath and storm.* **[The world against the Arabs of the south.]**

[15:35] They shall smite one upon another, and they shall smite down a great multitude of stars upon the earth, even their own star; **[Seems to be referring to meteorite or flying ships of some kind.]**

[15:38] And, after that, heavy storm clouds shall be stirred up from the south, and from the north, and another part from the west. the tempest that was to cause destruction shall rise, to destroy all the earth and its inhabitants, and shall pour out upon every high and lofty place a terrible tempest. **[The war is nothing compared to the "natural destructions."]**

Esdras Again

If we look at another of the Esdras books we get even more details. The details of the horrendous life humans have during the pre-tribulation war that precedes the Tribulation plagues are probably best presented in this second book of Esdras. The following excerpts need no discussion and come from chapters 12 through 17. I have eliminated much of the gruesome details as unnecessary for this work, but all should read the book to increase their awareness of this terrible time. The description includes wars, earthquakes, fire, famines, massive slaughter, judgment for the wicked, a short respite, little hope for all that remain, the final battle, and total destruction of the Earth.

Moslems Take Control of Europe

The days are coming *that the Earth will be under an empire more terrible than any before.*

It will be ruled by 12 kings. *One after another. The 2nd to come to the throne will have the longest reign of the 12.* **[I know Revelation indicated 10 nations take over the Christian world, but you could see from the previous list of Moslem states that it could be interpreted as more.]**

After the 2nd king's rule, *great conflicts will arise and will bring the empire into danger of falling, yet it will not fall then, but will be restored to its original strength.* **[The Moslem dominated kingdom will have some shaky times, but will not fall quickly.]**

Moslems Lose Control

Eight trivial kings *follow. Two will be left until the end itself. In the last years of the empire, the most high will bring to the throne 3 kings who will restore much of its strength and rule over the Earth more oppressively than any before.* **[Whether 2 kings, as indicated in Nostradamus and Mother Shipton's works, or three, as**

implied here; the will Moslem's lose control and a new evil emerges worse than the previous rulers.]

All the nations will leave their territories and unite in a countless host with a common intent, to **wage war against God**. *Then, my son will destroy them without effort and with the lake of fire.* [**This is identical to the Nostradamus description and the Revelation description of the pre-tribulation War.**]

After the time of famine everywhere, *Alas for the world and its inhabitants! The sword that will destroy them is not far away.* [**Soon after the last Dark Age will be the war.**]

Moslem Revelation

The Koran has a somewhat different prophecy concerning the years to come. In their prophecy, the Moslems taking over the world appear as an attempt to convert people to the truth while the pushing back of the Moslems is depicted only as the beast destroying the good. The section is known as "The Final Signs of Islam".

Comet Strike

The ground will cave in: one in the east, one in the west, and one in Hejaz, Saudi Arabia. **[It doesn't really say what is making the indentions, but from other works it appears to be meteors.]**

Pre-tribulation Dark Age

__Fog or smoke will cover__ the skies for forty days. **[The beginning of the pre-Tribulation Dark Age. This 40 days and 40 years continues to come up in these predictions.]**

__The nonbelievers__ will fall unconscious, while Muslims will be ill. **[Plague predicted.]**

__The skies will then clear up__. A night three nights long will follow the fog. Then the sun will rise in the west. **[There is no mistaking this prediction. A major earth shift is clearly identified here.]**

Moslems Take Control

People's repentance will not be accepted after this incident.
[This could be a reference to the Moslem hordes trying to take over Europe.]

Antichrist Gains Acceptance

The Beast from the earth will miraculously emerge from Mount Safaa in Makkah, causing a split in the ground. **[The beast emerges and pushes the Moslems back to Mount Safaa.]**

Mark of the Beast

<u>The beast will be able to talk</u> to people and mark the faces of people, making the believers' faces glitter, and the nonbelievers' faces darkened. **[The beast takes over the world and the mark of the beast is noted here.]**

<u>A breeze from the south causes</u> sores in the armpits of Muslims, which they will die of as a result. **[Another plague.]**

<u>The Ka'aba will be destroyed</u> by non-Muslim African group. Kufr will be rampant. Haj will be discontinued. The Qur'an will be lifted from the heart of the people, 30 years after the ruler Muquad's death. **[The Moslems do not have a good position under the world rule of the beast.]**

The fire will follow people to Syria, after which it will stop.

201

Greek Revelations

"Sibylline Oracles"

The Greek writers provided a great deal of confirmation and detail to our imminent future. The "Sibylline Oracles" is a long thesis written about 18 hundred years ago and covers many, many elements of our life. Some are not too well thought out. This one seems to be right on the money.

Comet Strike

2-235 His chariot celestial, and on earth arriving, shall to all the world display three evil signs **[This seems to be talking about a comet strike, which is eluded to in many prophecies.]**

2-245 A mighty stream of burning fire from heaven, And every place consume, earth, ocean vast, gleaming sea, and the heavenly sky; and heavenly lights shall break up into one, and into outward form all-desolate. For stars from heaven shall fall into all seas. **[The comet is predicted.]**

Pre-Tribulation War

2-410-And desolations shall thy cities be and in the west there shall a star shine forth which they will call a comet, sign to men of the sword and of famine and of death, and murder of great **[The falling comet signals the Moslem horde to the pre-tribulation war.]**

Moslems Take Control

2-435 *Of Asia, even thrice as many goods shall Asia back again from Rome receive, and her destructive outrage pay her back. As many as from Asia ever served a house of the Italians, twenty times as many Italians shall in Asia serve.* **[During the War, the Moslems imposed three times as much casualty and tribute as the earlier Roman Empire.]**

Moslems are Pushed Back

2-483-*In later generations into Asia's prosperous land shall come a* **man unheard of unjust**, *fiery; And this man wields the thunderbolt and all Asia shall sustain an evil yoke, and her soil shall drink much murder.* **[The Moslems are finally pushed out of Europe and the world is taken over by the "man of unheard of unjust", a man that Nostradamus called the Prince of Hell.]**

Gnostic Revelations

"Origin of the World"-
This Book references the darkness, War, major problem with the Oceans, meteors, destruction of Heaven and earth, the antichrist and many of the other components of the Biblical prophecy.

Comet Strike

Before the consummation of the age, the whole place will shake with great thundering. **[Just like the Biblical prophecy, the Earth will have huge Earthquakes possibly from the comet strike predicted by others.]**

Pre-Tribulation War

Then the kings will be intoxicated with the fiery sword, and they will wage war against one another. **[Just like the Biblical prophecy, huge wars will occur before the Tribulation period.]**

Pre-tribulation Dark Age

Then the seas will be disturbed by those wars. **[Just like the Biblical prophecy, something terrible will happen to the oceans just before or during the great tribulation Dark Age.]**

Then the sun will become dark, and the moon will cause its light to cease. The stars of the sky will cancel their circuits. And a great clap of thunder will come out of a great force that is above all the forces of chaos. **[This may be the indication of great meteor showers which cause the sky to be darkened just like past meteor bombardments.]**

"Our Great Power"-

At first our future according to "The Concept of Our Great Power" seems to be somewhat different than the Revelation and Nostradamus stories as shown in the graphic below, but let's look a little closer and look at the similarities.

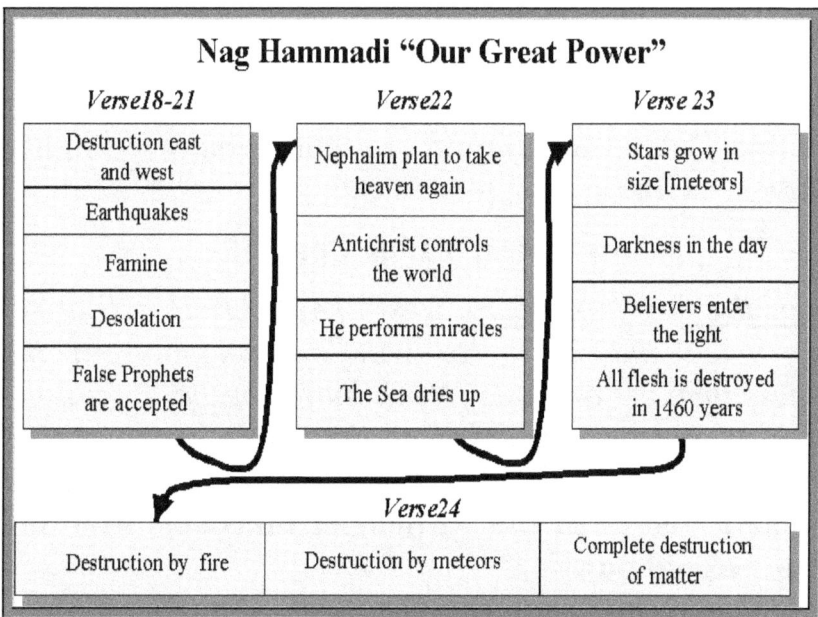

Comet Strike

[18-20] He [God] made the border of the West desolate, and destroyed the East. Then the wrath of the archons burned, the cities were destroyed; and the mountains dissolved. "The archon" [Major heavenly body] came along with the archons [lesser heavenly bodies] from the western region, to the East. **[Assuming archons means heavenly bodies, the writer is apparently talking about the great comet attack that comes from west to east according to Nostradamus and other writings.]**

Pre-Tribulation Dark Age

[21]Then the earth trembled, *and the cities were troubled.* *[Earthquakes after the comet]*

[21]Moreover, the birds *ate and were filled with their dead.* *[Famine after the comet.]*

Pre-Tribulation War

[21]Then when the [comet] times were completed, then "wickedness" [Moslem Horde] arose mightily even until the final end of the Age. **[Probably identifying the pre-Tribulation War, but could be addressing "mighty wickedness" without war.]**

Antichrist Takes Control

[22]Then the "archon of the west" *[Antichrist] will arise, and approach the East. He will instruct men in his wickedness then the archons [Nephilim] sent the "imitator" to that man and he reigned over the whole earth.* **[This "archon" could be heavenly body, but it makes more sense that it refers to the Antichrist takeover after the Moslems pushed out.]**

[23]When he [Antichrist] has completed *the established time of the kingdom of the earth, then the cleansing of the souls will come.* **[Tribulation follows the antichrist abomination.]**

When Is The End?

For completeness, let's look at the END OF THE EARTH!

After the Comet strike, a shift in the Earth, an Ice Age or terrible hunger, the suffering, the Moslem War, and the push back by the rest of the world, there is still more as finally, God Incarnate comes backs takes away his people, there is a long period of peace followed by a final war that ends in destruction. We are given dates of the end of time by several people; Nostradamus, Mother Shipton, and Daniel. Of these Nostradamus is the most direct, however, Daniels's descriptions tie in all of the major events and provides great confirmation. Mother Shipton's date is also very matter-of-fact. The end will be about 3800AD.

Nostradamus End

Nostradamus clearly stated, in plain French that his prophecies would extend to the year **3797AD.**

Mother Shipton End

"When the world to an end shall come, in eighteen hundred and eighty one." **[Because everything she wrote used the year 2026 as the starting point, from this we find that the end will be the year 3907.]**

Daniel's End

207

According this interpretation of Daniel's words, the Millennium Reign will end around **3850AD** and the Tribulation period will end around 2500AD. A reasonable assumption concerning all the "new prophecies" is that they will take another 500 years which indicates that –

The fun will begin around 2030 so get ready.

---

*Daniel 12:7-It shall be for a time, times and a half: [**2,500 time periods-(years)**] and when they have made an end breaking in pieces the power of the holy people, all these things shall be finished- from the time that the burnt offering shall be taken away and the "abomination that makes desolate" [antichrist] is set up, there shall be 1,290 days Blessed is he that waiteth and cometh to the 1,335 days, but go thy way until the end of the days.* [**This verse may indicate that there will probably be 2,500 years from the time of Daniel until the end of the pre-Tribulation War. That puts the comet coming at any time. If we assume the remaining descriptions are years rather than days, it seems to indicate that the end will be approximately 3850 AD.**]

Daniel 12:11- From the time that the continual burnt-offering shall be taken away, and the "abomination that maketh desolate" set up; there shall be a thousand and two hundred and ninety days. [**If we assume that the "abomination that makes desolate" is the time when Nostradamus' "Prince of Hell" rules, and the time periods are actually years rather than days, then we are talking about 1290 years before 3805 or 2515 AD If we assume that all of the events leading up to the 3.5 year tribulation period take another 500 years, then the Daniel, Estras, Revelation, Nostradamus and Mother Shipton prophecies will begin very soon.**]

208

Daniel 12:12- *Blessed is he that waiteth, and cometh to the thousand three hundred and five and thirty days. But go thou thy way till the end be; for thou shalt rest, and shalt stand in thy lot, at the end of the days.* **[This could probably give us an idea about how long the final battles last, about 45 years, and then a new earth is created about 3850 AD.]**

Conclusions

Hopefully you got something out of my rantings here. While I am biased toward not believing people when they talk about gloom and doom and get wealthy from it, but I tried to show both sides of this important topic.

On the "**man is killing the earth**" side we find the following"

- Many in the science community have endorsed the Manmade Global Warming issue

- Charts showing a slight temperature increase are easily found

- Charts showing slight increases in water height can be found

- Charts showing massive increases in CO_2 are easily found

- Charts showing slight increases in NO_2 and CH_4 can be found

- Our leaders including our President have embraced the possibility.

- Our congress has provided billions in guaranteed funds, price offsets and incentives, grants to allow easy Green industry expansion.

- Our government is forcing the elimination of Coal use even though it will cost the US a fortune to convert to much less productive energy sources.

- EMAIL traffic obtained with FOIA orders showed massive manipulation of temperature data to make it look like the temperatures were increasing. Even after the details were brought out, the temperature charts were not fixed by NOAA. Showing NOAA didn't need Satellite data, Buoy data, or historic data to have a gut fell that man's civilization is killing our planet.

- Climatologists have made almost 100 temperature tracking and plotting models and all show the Earth will soon be a dead planet. [OK! All the models have failed to predict the temperature at all, but they are being used to prove that we should quite driving cars.

On the "**Possible Ice Age**" side we find the following"

- Many in the science community have indicated they were wrong about their assumptions concerning Manmade Global Warming

- Charts showing a slight temperature decreases are easily found

- Charts showing slight decreases in water height can be found

- Charts showing less CO2 increase than at earlier times can be seen.

- Charts showing increases in NO2 and CH4 much lower than at previous times can be found

- Our leaders including V.P. Al. Gore have made fortunes pushing Green industries and pushing Congress to shore up the industries artificially while they reap the benefits.

- Huge hydrocarbon producing nuclear wars have no caused global warming and neither did dinosaur flatulence.

- The temperature cycles and sun irradiance cycles seem to match even though man does not control the sun.

- In the past warming periods of about 10 thousand years always accompanied the end of a Glacial Period. It has been 10 thousand years since the Pleistocene extinction that ended the last glacial Period.

There is absolutely no question that the NOAA scientists know that CO_2 does not leave the air for thousands of years. While they reported to the public this was a fact, the same scientists took aerosol CO_2 reading and added them to the Ice-Core CO_2 level data to make more fear.

Canada or Florida

Please do not go out and purchase northern Canada land to make a fortune as fear of the upcoming calamity drives anxiety, but buying land in Florida or even Mexico might be a smart move as the Ice Age is certainly coming. I know your solar heaters are going to be of little comfort as the reduced use of coal forces the price of electricity through the roof

Investments

It should be known that Al Gore has been moving his money away for the volatile green industries to other more reasonable companies like "Apple" so that when the bubble hits his empire will not be affected. It will not take many more years before the NOAA will have to come clean to the American people as temperature rise is becoming more and more difficult to manufacture.

About the Author

Steve Preston is a long time author of scientific, esoteric facts. His series on the creation of mankind is shown below. The series focuses on the painful truths rather than whitewashed details that make us comfortable. If you are interested in the truth instead of comfort, please continue to read and, while you are at it, review other works by Mr. Preston as shown below. Like this one on global warming errors, most of the books are not politically correct and most have readers who hate the same books that other truly love. He is only interested in finding the truth rather than making people feel good about what they thought was truth for years.

Eight Part Series "History of Mankind"
The First Creation of Man
The Second Creation of Man
The Creation Of Adam And Eve
The Antediluvian War Years
Man After the Flood
A Closer Look At Ancient History
A New View Of Modern History
The 20th Century To The End Of Time

Other Works

Kingdoms Before the Flood
Egyptians In Ireland
Behind the Tower of Babel
When Giants Ruled the Earth
Ancient History of Flying
Lizard People
Who Really Discovered the Americas?
America's Civil War Lie
Living On Mars, Venus, and the Moon
Disgusting Display
Vibrational Matter
Adam's First Wife
Allah' God of the Moon
Anakim Gods
Closer Look At Genesis
Walk Through Time or a Wall
Moses Saved Egypt
Biophotonics and Healing
The Devil
Mystery of Photons and Light
Mysterious Pyramids
Fast History of MILES Training
World War Zero
Why the King James Bible Failed
Anthropic Reality
Victory of the Earth
Races of Men
Our 10-Dimensional Universe
Four Armageddons
Not from Space
Creation and Death of Dinosaurs
Self-Soul Spirit
Driven Underground
God Didn't Make The Ape

The Antichrist
DNA of Our Ancestors
Strange, Powerful & Dangerous Women
History Confirmed By The Bible
Mysteries of the Exodus
Complex Earth
World War Before
Meaning of Life and Light
The Book Of Odd
Why Are There So Many Anomalies?
Releasing Your Consciousness
Errors in Understanding
Retiming the Earth
Great American Quiz
Stupid Science
Our Very Odd Presidents
Sex Crazed Angels
World War with Heaven
More Oddness
Closer look at Photons
The Bad Side of Lincoln
New look at the Bible
Life Resonance
Truth About Vampires
Why Rome Fought the Berserkers
Awaken the Departed
Changes of the King James Bible
Six Deaths of Man
What about Time Travel?
Who Made the Pyramids
American School Disaster
Genesis Companion
Slip Through a Wall
Today's Monsters

Investigations In Various Countries

As a serious speculation and history writer, the author has investigated in Israel. He is the guy on the left standing on an embankment in the Negev Desert where the Dead Sea Scrolls were found. The image to the right is of the author on the Acropolis investigating the statues of antiquity there.

Next, the author rides a camel to the great pyramid of Egypt and travels in a New Zealand cave home of the ancient Maori.

While this book required mostly investigation of common sense. The other places were helpful in showing him how amazing things can be.

Thanks for Reading!